Edible Insects and Human Evolution

UNIVERSITY PRESS OF FLORIDA

Florida A&M University, Tallahassee
Florida Atlantic University, Boca Raton
Florida Gulf Coast University, Ft. Myers
Florida International University, Miami
Florida State University, Tallahassee
New College of Florida, Sarasota
University of Central Florida, Orlando
University of Florida, Gainesville
University of North Florida, Jacksonville
University of South Florida, Tampa
University of West Florida, Pensacola

Edible Insects and Human Evolution

JULIE J. LESNIK

University Press of Florida
Gainesville · Tallahassee · Tampa · Boca Raton
Pensacola · Orlando · Miami · Jacksonville · Ft. Myers · Sarasota

Printed in the United States of America on acid-free paper

This book may be available in an electronic edition.

23 22 21 20 19 18 6 5 4 3 2 1

Library of Congress Cataloging-in-Publication Data
Names: Lesnik, Julie J., author.
Title: Edible insects and human evolution / Julie J. Lesnik.
Description: Gainesville : University Press of Florida, 2018. | Includes bibliographical references and index.
Identifiers: LCCN 2017055302 | ISBN 9780813056999 (cloth : alk. paper)
Subjects: LCSH: Entomophagy. | Edible insects. | Food habits. | Human evolution.
Classification: LCC GN409.5 .L47 2018 | DDC 394.1/2—dc23
LC record available at https:// lccn.loc.gov_2017055302

The University Press of Florida is the scholarly publishing agency for the State University System of Florida, comprising Florida A&M University, Florida Atlantic University, Florida Gulf Coast University, Florida International University, Florida State University, New College of Florida, University of Central Florida, University of Florida, University of North Florida, University of South Florida, and University of West Florida.

University Press of Florida
15 Northwest 15th Street
Gainesville, FL 32611-2079
http://upress.ufl.edu

For Uco and his dad

CONTENTS

FIGURES

MAPS

TABLES

PREFACE

Reconstructing the role of insects in the diet of our hominin ancestors requires the integration of many academic subdisciplines. Edible insects are essentially invisible in the archaeological record of our early ancestors, so in order to assess their nutritional benefits, many lines of reasoning must be used in tandem to create and test models. This book is testament to the benefits of interdisciplinary research.

I regularly speak on edible insects at anthropology conferences, but over the past three years I have been presenting my work to scholars in other disciplines, including entomology and food science. This experience has inspired me to write this book as a narrative that will introduce biological anthropology and the study of human evolution to scholars outside the field of anthropology. Ever-growing numbers of people are interested in insects as food, and by writing this book for them, I get to promote a better understanding of human evolution.

For students taking undergraduate anthropology classes, this book also serves as a good introduction to the topics of biological anthropology, archaeology, food anthropology, historical anthropology, and anthropology as activism. Through this project I hope to promote the benefits of cross-discipline study and the holism of the field of anthropology to the next generation of anthropologists. Furthermore, I hope that graduate students interested in the evolution of the human diet will find this book useful when they set up their own research projects. We have much to learn about what our ancestors ate, and I identify numerous methods that are available to us to address such questions.

Overall, my goal is to represent edible insects in the context of the elegant processes that make us human and within the context of the multitude of factors that affect our choice of food. To accomplish this, I decided to commit to the following principles:

1) I would avoid portraying edible insects using fear- or disgust-triggering images or language. Insects are a nutritious food source with great value in cultures around the world and with great potential in our own culture. I want this book to demonstrate their value, not their novelty.
2) I would do my best to contribute to a decolonized anthropology. There is a notable separation in consumption of and attitudes toward edible insects between cultures that can be identified as western and nonwestern. Decolonizing anthropology requires us to acknowledge that the differences in these cultures may be less about geography, economics, or politics and more about colonial history. The discipline of anthropology was born in the context of the differing power dynamics of these groups. For many decades, western researchers have placed themselves in nonwestern cultures to learn more about their ways. I aim to acknowledge the unique identities of the cultures that eat insects but also show how they have similarities in their histories because they are on the same side of the global divide.
3) I would highlight the important role of foraging over the course of human evolution, and in doing so acknowledge the need for feminist perspectives in paleoanthropology that will move beyond using traditional modern gender roles as a basis for interpreting hominin behaviors in the past. I seek to build on the dialog started by trailblazers such as Sarah Hrdy and Adrienne Zihlman, who have rigorously pursued this goal for decades, so that one day representing gender fairly will become imbedded in the framework of paleoanthropological inquiry.

Having said all that, there is really only one simple message: Insects are food. They occupy an important role in the food web. They sustain numerous insect specialists, including species of birds, amphibians, and mammals, and are incorporated into the diets of countless animals who consume a variety of foods. One species in particular is known to eat insects in varying degrees across its wide geographic range: humans. The human diet contains insects and this book contemplates both the origin and the future of this aspect of the human condition.

ACKNOWLEDGMENTS

This book could not exist if I had not received abundant support from many different people. Any shortcomings in the manuscript are my own, but anything positive is due to the collected efforts of those mentioned below.

I never would have imagined writing a book had it not been for my friend and mentor John Hawks. At the 2013 AAPA meetings in Knoxville, John casually suggested "you should write a book," and later that afternoon I was speaking with potential publishers. I found making the pitch exhilarating, and that energy propelled me through the next couple of years of my career, ultimately helping me transition from adjunct teaching to a tenure-track position.

Milford Wolpoff, was the first to suggest that I study termites as a hominin food source. He recognized my interest in behavioral models and my quirky desire to study something slightly outside the norm. I never imagined that my research would lead me here, but I expect that he did.

I experienced an outpouring of generosity from many people around the world for this research. I thank Francis Thackeray, especially, who drove me to dig my first termite mound and made me realize I could be a professional in this field. This project would not exist if Francis had not taken me on as a visiting researcher in 2006.

Lucinda Backwell has kind enough to help me in countless ways over the years, from giving me advice as a student to letting me sit in her lab for weeks as I worked on this manuscript. Most important, I would not have had a research project at all if she had not included termite digging in her bone tool experiments. In terms of my scholarship, I stand on her shoulders.

Davorka Radovčić provided me a home away from home so that I could focus on writing this book. I am so lucky to have such a selfless and encouraging friend.

Francis Duncan and Shandu Netshifhefhe generously invited me to join them on their research project in Limpopo, South Africa. I am grateful that they introduced me to the women who collect and sell termites in Thohoyandou. That experience enriched this book significantly.

My position at Wayne State University has been a blessing throughout this process. The Department of Anthropology and the entire College of Liberal Arts and Sciences have been supportive of my work in countless ways. I am most grateful to have Stephen Chrisomalis as my faculty mentor. Steve went above and beyond in helping me with this book. He read my first draft of nonsense and was able to see the potential in the project. He then kept tabs on me every step of the way. Without his help, I never would have pulled this together.

Students at Wayne State have also been amazing. I am grateful to my research assistants Nicole Lopinski, Jasna Cakmak, and John Anderson.

I am indebted to Margaret Schoeninger, Bill McGrew, and Alan Mann for their tremendous support and thoughtful comments on an earlier version of this manuscript. With their help I was able to actualize the vision I had for this book. I also want to thank Nina Jablonski, whose voice was in my head throughout this process. She kindly met with me to talk about taking on a book project, and her book *Skin: A Natural History* was always by my side for inspiration.

There is no way that I could fully express my gratitude to my acquisition editor, Meredith Babb, or to my copyeditor, Kate Babbitt. Although I know they were just doing their jobs, they never made it feel that way. I always felt that they cared about me and truly believed in this project. I cannot imagine accomplishing this with any other editors.

I am fortunate to have friends who helped me in various ways, from reading parts of my manuscript to answering random questions over email. Thank you especially to Rob O'Malley, Clay Magill, David George Gordon, Dave Gracer, Caroline van Sickle, Michaela Howells, Christopher Lynn, Elizabeth Rowe, Cara Ocobock, Ann Laffey, and Amanda Logan.

My brother Jeff Lesnik helped me edit my favorite photo—Gombe chimpanzee Glitter nursing her daughter while fishing for termites. This moment captures the heart of my book, and Jeff's professional eye helped me portray it here in black and white.

I also thank my husband, Charlie Klecha, for his unwavering love and support throughout this whole process. His #gojuliego encouragement was often all I needed to buckle down and write just a little more.

Finally, I thank the American Association of University Women for funding this work. I am honored that this important organization found my project worthy of support.

1

Introduction to Entomophagy Anthropology

As a lifelong picky eater, I never imagined that I would become a promoter of the benefits of insects in our modern diet. My study of edible insects started out as a purely scholarly pursuit. I came into the field of paleoanthropology with a desire to study the evolution of tool use. To me, the transition to more sophisticated and complex ways of exploiting the environment represented a critical stage in human evolution. Our ancestors that used tools had access to increased resources, and they could support the nutritional demands of their increasing brain sizes. As brain size grew, more innovation could occur and more resources could be procured. I wanted to find a project that would enable me to investigate the relationship between tools, diet, and brain size. Ultimately, and unexpectedly, I found my research focus not on tools but instead on the insects that became available through tool use.

My research has focused primarily on the dietary contributions of termites. Our closest living primate cousins, the chimpanzees, are notable termite foragers; they make tools from sticks, leaves, or grass to penetrate open passageways of termite mounds and "fish" out termites. People today also consume termites in tropical regions across the continents, using more sophisticated technology than chimpanzees but still relying on nonmechanical means. Because both humans and chimpanzees eat termites, it seems likely that our last common ancestor and its descendants on both sides also ate termites.

The descendants along the human lineage are collectively known as hominins and include humans, our direct ancestors, and closely related extinct taxa.[1] Many paleoanthropologists have noted that early hominins would have been capable of using simple tools to extract social insects from their nests in ways similar to chimpanzees today (e.g., Bartholomew and Birdsell 1953; Mann 1972). Panger and colleagues (2002) describe the rationale:

> Because most early hominins are associated with nonarid environments such as riparian forests, densely wooded habitats, and mosaic habitats with some grasslands, we assume that such high-energy, difficult-to-access foods such as nuts, social insects, and honey were available. We further assume that raw materials for making tools, among them sticks, grasses, leaves, and appropriate stones, were also available to all early hominins. There is little doubt that hominins, given a body mass comparable to or greater than that of chimpanzees, had the strength to perform a variety of tool-using activities, including the manufacture and use of stone tools. (Capuchins, which, at 2 to 3 kg, are about the size of a house cat, are able to make stone tools.) Thus, we assume that the ecological impetus, raw materials, and necessary strength for tool use were available to early hominins from the time of the split between hominins and panins [chimpanzees and bonobos]. (236)

Working from these same assumptions, my research goals are to reconstruct the dietary and behavioral components of insect foods for hominin life. I have concluded that termites and other insects were a reliable and valuable resource that hominins likely benefited from. What is particularly notable about insects is that they are nutrient dense and that it is much simpler to acquire them than it is to obtain other similarly nutritious resources. Insects are an animal food; they provide many of the same benefits as meat, eggs, and milk. In our society, these products come from domesticated animals, but in nature, these resources are difficult to obtain. Yet insects are abundant; although difficult to calculate, it is suspected that they make up the largest proportion of the earth's terrestrial fauna biomass.[2] Although many insects are secretive, social insects such as termites, ants, and bees are especially easy to find because they live in groups of thousands or even millions. Relying on the capture of large game is a risky way to meet dietary demands, especially when many of the same benefits can be received by visiting a termite nest for a short period of time. Insects thus provide an appealing resource option for individuals who cannot afford to take risks in how they procure food. Although some hominin individuals may not have been affected when they had to endure short periods without game, female hominins, who for much of their adult lives were responsible for meeting their own nutritional demands plus the needs of a gestating fetus or nursing infant, would have benefited the most from the dependability of insect foods.

After detailing the numerous benefits of edible insects in my research, the lack of this food in my own culture became glaringly obvious. I began to wonder why we don't eat insects in the United States. Because I believe insects provided an important food source over the course of human evolution, I decided that the question could best be addressed by reframing it: Why did people in some cultures stop eating insects? As I began to investigate this new line of research I was introduced to an ever-growing community of edible insect enthusiasts who promote the normalization of this vastly underutilized resource. Some of these advocates are even promoting insects as part of the paleo diet craze because they fit the grain- and dairy-free requirements of this trend. As a researcher of insects in real paleo diets, I now aim to apply my work to real-life problems of food security and sustainability. In this book, I will demonstrate how theory from the field of biological anthropology adds to the written record on the benefits of edible insects and how the academic pursuit of the true paleo diet provides convincing evidence of the importance of insects as a food source.

Entomophagy and Other Terms for Eating Insects

Carnivores (meat eaters), herbivores (plant eaters), and omnivores (those who have a mixed diet): these definitions are familiar to most people thanks in part to abundant educational media directed at children who are captivated by dinosaurs. However, in biology, there are numerous variations on these themes. These include obligate diets, which means the entire diet comes from a resource, and facultative forms, where the resource makes up the majority of the diet, but it is supplemented with other food items. As a species, humans are omnivores. Our bodies require nutrients that in nature come from both plant and animal foods. However, the human condition enables people to be successful with highly variable proportions of these resources. It is possible for people to be facultative carnivores or facultative herbivores as long as all dietary nutrients are provided in some way. In modern society, these nutrients may come from supplements or enriched foods, but in nature, people have found ingenious ways of extracting essential nutrients from diverse habitats.

Another term is insectivore, which indicates a specialization in eating insects. Across the animal kingdom, species of invertebrates, fish, amphibians, birds, lizards, and mammals make a living by eating insects. In mammals, the word insectivore often indicated placement within the now-defunct taxonomic order Insectivora. This category was a catch-all taxon that included primitive-looking, small, insectivorous fossil mammals because they were thought to resemble the ancestral stock from which all placental mammals arose. Extant taxa of the order included shrews, moles, and hedgehogs, but advances in genetics have determined that these species do not share a recent common origin. This has rendered the order Insectivora taxonomically meaningless, and its former constituents have been placed into other orders (Stanhope et al. 1998).

Additional confusion regarding the term insectivore arises when its derived form, insectivory, is used. Unlike insectivore, which indicates either obligate or facultative specialization on the resource, insectivory is used more broadly to indicate consumption of insects in any quantity. For instance, there are few insectivorous species in the order Primates, but primatologists often record the rates of insectivory for species that specialize on other foods such as fruit and supplement their diets with only small amounts of insects.

A synonym for insectivory is entomophagy, but this word also has a complicated meaning and usage. One benefit of the word entomophagy is that unlike insectivory, it does not imply specialization (phagus = eating; vore = one that feeds). Even people who eat insects every day are not facultative insectivores, since the majority of their diets is composed of other resources. Thus, entomophagy is a more accurate term for describing occasional or supplemental insect eating. This term is more commonly used than insectivory in scholarly literature pertaining to insects as human food. However, the "phagy" part of the term implies an infrequency that can suggest that the occasional use of the food is an exceptional or even an abhorrent behavior. There is no comparable terminology for eating other sources of animal protein—for instance, bovinaphagy is not a word to indicate beef consumption, even though the meat portions of diets vary greatly by culture and some animals are used only occasionally or not at all. Examples of commonly used terms that include "phagy" are geophagy, the consumption of soil, and coprophagy, the consumption of feces, both of which, when applied to people, indicate feeding on resources that are essentially nonnutritive to humans and thus behavior that demands an explanation such as a severe mineral deficiency or an eating disorder. Thus, the term entomophagy may incorrectly imply that insects are not a high-quality food for people (Evans et al. 2015).

A problem with these terms that is unique to this book is using the words commonly used in academic disciplines to talk about insect consumption over the course of human evolution. If entomophagy is the term that is used most often when describing the consumption of insects in the present-day human diet and insectivory is the term most commonly used to describe the insect portion of the nonhuman primate diet, what word should be used for the insect eating our hominin ancestors did? The two words are technically synonyms, but in this case, choosing one word over the other may imply further meaning regarding the human condition. Thus, for most of my writing, you will find that I avoid specialized vocabulary and discuss the acts of foraging for and consuming insects just as I would write about any food source, whether it was meat or mushrooms.

The Biggest Champions of Insect Consumption

Academic interest in edible insects reflects trends in our society. In recent years, there has been an increase in research about the use and benefits of insects, but before 2010, the topic was largely understudied. While the main goal of this book is to focus on the evolutionary and anthropological elements that are important for understanding the use of edible insects, a brief history of how this topic has become increasingly pertinent to food culture in our society is a good way to begin.

In 1885, Vincent Holt published a small pamphlet in London that touted the benefits of entomophagy titled *Why Not Eat Insects*? It is still in print after all these years, but unfortunately it is probably seen as an eccentric novelty item and is probably not taken seriously as thought-provoking discourse. Holt provided evidence of how other cultures prepared and ate insects in the late nineteenth century and claimed that the nutritional offerings of insects could aid the Victorian working man. However, he knew the difficulty he faced in convincing his readers of the benefits of insects as food:

> In entering upon this work I am fully conscious of the difficulty of battling against a long-existing and deep-rooted public prejudice. I only ask of my readers a fair hearing, an impartial consideration of my arguments, and an unbiased judgment. If these be granted, I feel sure that many will be persuaded to make practical proof of the expediency of using insects as food. There are insects and insects. My insects are all vegetable feeders, clean, palatable, wholesome, and decidedly more particular in their feeding than ourselves. While I am confident that they will never condescend to eat us, I am equally confident that, on finding out how good they are, we shall some day right gladly cook and eat them. (Holt 1885, 5–6)

Maybe Holt's "some day" is today. Holt's insights hold up after all this time, and with the help of other champions who have come along since the publication of his pamphlet, real momentum may be building toward a more widespread acceptance of the idea of edible insects. In 1951, entomologist F. S. Bodenheimer published a book titled *Insects as Human Food: A Chapter of the Ecology of Man*. It is still the most comprehensive account of different edible-insect practices around the globe, but like Holt, Bodenheimer did not change the public's attitude toward eating insects. The book was written with the goals of investigating a topic that provoked curiosity and discussing the nutritional importance of insects for "primitive" man. By "primitive," Bodenheimer did not mean our human ancestors; he meant it in the stereotypically negative way of implying that nonwestern cultures were uncivilized cultures. These perceptions run deep in European history and are still pervasive today. Even though his descriptions detailed how delightful insects were as food in local cultures, Bodenheimer unwittingly perpetuated the persistence of the stigma by portraying the foods as something only "Others" eat.

In 1975, entomologist Gene DeFoliart began compiling references related to insects as food with the goal of publishing an updated version of Bodenheimer's book. His project turned into an outreach effort that included the founding of *The Food Insects Newsletter*, multiple articles that include "Insects as Food: Why the Western Attitude Is Important," and the publication of an extensive bibliography that he offered for free on his personal website. One of DeFoliart's main messages was that the bias in the "west" against insects was causing a gradual reduction of their use as food worldwide. Elimination of this highly nutritious food source without a suitable replacement is especially problematic in areas where people find it difficult to meet their nutritional needs. Because of his passion, his cultural sensitivity, and his early adoption of open online access, DeFoliart opened many entomologists' minds to the idea of edible insects.

Archaeologists and paleoanthropologists who reconstruct which foods were important in the past have commonly overlooked insects. However, there are some notable exceptions. Mark Sutton is an archaeologist who has worked in the Great Basin area of the United States (Sutton 1988). Through his research, he found that insects were important as food in the culture he was reconstructing, and he recognized that most archaeologists blindly ignore insects during excavation and in their analytical methods. Sutton (1995) suggested that insects found in archaeological contexts could provide valuable information about past ways of life.

Primatologist W. C. McGrew is an expert on chimpanzee tool use and insect foraging. Through his many years of research on the insect portion of the diet of our closest living relatives, he became convinced of the likelihood that our hominin ancestors also ate insects. He has consistently argued that insects are just as valuable as meat and that reconstructions of the hominin diet need to take the nutritional value of insect foods into consideration. He has referred to insect consumption as "the other faunivory" as a way of placing them on an equal footing with meat (McGrew 2001; 2014).

Biological anthropologist Darna Dufour is well known for her extensive research on insects as a food resource for indigenous peoples of the Amazon basin. Her research, which focused on biological and behavioral responses to food shortages, identified the importance of edible insects to the Tukanoan people of Amazonia and detailed their nutritional contributions and cultural value. Although her work could have presented insects as a fallback food that was consumed only when the higher-valued fish was unavailable, she instead demonstrated that insects are an important and valuable resource through a detailed analysis of their rich nutritional content (Dufour 1987). Her methodology set the standard for what data are necessary for capturing the role of edible insects in a local diet.

The biggest break for promoting edible insects came at the 2010 TEDGlobal conference, where Marcel Dicke, an ecological entomologist from the Netherlands, gave a talk with the same title as Holt's manifesto: "Why

Not Eat Insects?" Instead of pointing out that insects are nutritious and are consumed by people all over the world, Dicke added his ecological perspective to demonstrate that insects, more than any traditionally raised livestock, have the potential to be sustainably cultivated. This notion generated excitement and began changing how many people think about insect consumption. Since 2010, farms for raising crickets for human consumption have become a reality in the United States and across Europe and over 100 insect-food startup companies have launched worldwide (Bugsolutely 2018). One of the pioneer companies is Chapul, which makes a protein bar out of cricket powder. Chapul founder and CEO Pat Crowley credits Dicke's TED talk for giving him the idea, and Crowley himself has now given three TEDx talks to continue spreading the message about the value of edible insects in the human diet (Crowley 2014a, 2014b, 2015). Crowley has a background in hydrology and was particularly motivated by the low volume of water it takes to produce a kilogram of crickets compared to the volume it takes to produce a kilogram of any traditionally raised livestock. Estimates suggest it takes 22,000 liters of water to produce a kilogram of beef, 3,500 liters to produce a kilogram of pork, 2,300 liters to produce a kilogram of chicken, and less than 500 liters to produce a kilogram of crickets (Van Huis et al. 2013; Halloran et al. 2017).[3]

Long before cricket protein bars hit the market, David George Gordon, the "Bug Chef," was promoting edible insects with his *Eat-A-Bug Cookbook*, first published in 1998. Gordon's recipes highlight insects instead of hiding them and work with their natural flavors to make delicious dishes (figure 1.1). According to Gordon, "many bugs have subtle flavors, reminiscent of crab or shrimp, [while] a few are surprisingly flavorful. Prime examples are stink bugs, eaten with relish in Mexico, and giant water bugs, popular foods in Thai and Vietnamese homes" (personal communication, September 25, 2017). Gordon's recipe for giant water bugs balances their earthy taste with a subtly sweet ginger sauce. It is important to Gordon that people experience the insects as delicious, since that is the only way they will accept them as food.

In 2013, the Food and Agricultural Organization (FAO) of the United Nations published an extensive statement detailing the benefits of insects as food for people and as feed for livestock (Van Huis et al. 2013). The report provides the nutritional contributions of a few different insects and investigates the environmental impacts of the current food system.

Figure 1.1. Dishes created by the Bug Chef, David George Gordon, highlight the insects instead of hiding them, such as this ants-on-a-log snack featuring Changbai ants (*Polyrachis* spp.). Photo by C. M. Cassady.

The most data currently exist for farmed crickets, which require less land and water and emit significantly less greenhouse gases than other forms of livestock. While the FAO report emphasized the potential of insects, it noted that a lot of work still needs to be done. The publication suggests four areas of research : 1) further documentation of the nutritional values of insects in order to promote insects more efficiently as a healthy food source; 2) investigations of the sustainability of farming insects and comparisons of the environmental impacts of harvesting and farming insects with the impact of farming other animals; 3) clarification of the socioeconomic benefits that insect gathering and farming can offer, with a focus on improving the food security of the poorest people on the planet; and 4) development of a clear and comprehensive legal framework at national and international levels that can pave the way for more investment, leading toward the full development (from the household scale to the industrial scale) of production and trade in insect products for food and feed internationally. I would add two points to this list: 5) application of psychological concepts such as group decision making and social change theory as argued by Shockley and Dossey (2014) to our understanding of the consumer market; and 6) appreciation of the cultural value of various insects to people around the world, both in the past and today.

Anthropology of Food

The reasons people choose certain foods are extensive and exceedingly complex (Mintz and Du Bois 2002). For one thing, human culture is complex, and for that reason, it is hard to define. While early definitions focused on customs, laws, morals, and other readily catalogued traits (Tylor 1871), broader definitions such as Goodenough's (1957) are more readily applied across societies with varying structures. Goodenough wrote, "A society's culture consists of whatever it is one has to know or believe in order to operate in a manner acceptable to its members" (167). While culture can be adaptive, it is not always so. Thus, understanding why people do the things they do often requires us to integrate multiple perspectives. A good example comes from studying the consumption of beef in India.

Much less beef is consumed in India than in other countries. What we perceive in the United States as a high-quality food source is largely ignored by people who practice Hinduism in India, even in the leanest of times and even when cows are widely present. Frederick Simoons, the author of *Eat Not This Flesh: Food Avoidances in the Old World* (1961), explained that protecting cows is an essential tenet in Hindu faith, which is the predominant religion in India. Hindus see cows as sacred and protect them for religious reasons. Simoons illustrated this point in a 1979 paper with a quote from Mahatma Gandhi:

> The central fact of Hinduism is cow protection. Cow protection to me is one of the most wonderful phenomenona in human evolution. It takes the human being beyond his species. The cow to me means the entire subhuman world. Man through the cow is enjoined to realize his identity with all that lives. (Gandhi 1954:3)

In 1966, Marvin Harris suggested that the management of cattle in India can be explained with an ecological framework. He argued that avoiding beef can be understood as adaptive: cows provide better resources when they are kept alive, including milk and the manure that is used for cooking fuel. When a cow is killed for its meat, the meals are not worth the loss of the other resources that have enabled people to successfully persist in the rather harsh, dry environments of India. Simoons and colleagues (1979) found Harris's arguments to be misinformed and a poor representation of Hindu culture. What followed was a decades-long debate among anthropologists that is now known as the "sacred cow controversy" (Simoons et al. 1979). However, both sides make valuable contributions toward understanding the culture. While there may be an adaptive reason for why meat avoidance came into practice, if you asked any Indian individual to explain the reasons for their choice, they will tell you that it is related to deep religious beliefs and cultural practices. Both perspectives help to fully understand this particular food choice.

The ideal view of anthropology is that there are multiple approaches to studying humankind and that the information obtained in these multiple ways is what creates a holistic view of people that is greater than the sum of its parts. However, as we can see with the sacred cow controversy, researchers become specialized and it can be difficult to remember that we must combine all of the various sorts of knowledge into a much bigger picture. These different specializations generally fall into one of the four subfields of anthropology: sociocultural, linguistic, archaeological, and biological. In this book, I aim to represent broad anthropological views and make sure to never discount the power of culture or cultural bias. However, I am a biological anthropologist, and this book is largely rooted in the theory and practice of that subdiscipline. Biological anthropology is the study of evolution as it pertains to the human species. Scholars approach this subject in numerous ways, but of most importance here are paleoanthropology (the study of hominin fossils), primatology (the study of our closest living relatives, the nonhuman primates), and human ecology (the study of how people living today interact with their environments).

Paleoanthropology, the study of our hominin ancestors, is characterized by the study of the bones and artifacts in the fossil and archaeological records. However, these materials are meaningful only when they are studied in a larger context, so paleoanthropologists rely on research in comparative anatomy, human ecology, and primatology in order to make their interpretations.

Human ecology research has benefited from the study of modern populations of hunter-gatherers. Today few populations rely entirely on these traditional ways of life, and much of what we know comes from studies conducted prior to the 1970s. In the contexts those earlier studies were based on, a person's ability to survive and reproduce was directly related to how well he or she navigated the challenges of their environment. In contrast, people living in industrial food systems have removed themselves from natural environmental pressures. For them, food is available year-round at the local grocery store and access to these foods is related to economics rather than the environment.

Studies of modern hunter-gatherers help scholars create analogies of how intelligent hominins navigate complex environments that are useful for reconstructing prehistoric ways of life. People in these populations of modern hunter-gatherers are no different anatomically or intellectually from any modern population, but unlike people in industrialized societies, they face pressures similar to those our early modern human ancestors experienced. By studying how modern populations navigate natural environments, we may be able to tease apart which of our present-day behaviors are related to the modernity of our species and which are related to our modern food systems.

While studies of hunter-gatherers are useful for reconstructing the behaviors of members of the genus *Homo*, for our earlier ancestors, paleoanthropologists often look to studies in primatology. Primatologists detail how our cousins in the order Primates interact socially and how that affects their ability to find food, secure mates, and successfully reproduce. Like us, nonhuman primates have relatively large brains compared to other mammal species and invest heavily in raising their offspring, two very important traits that create unique challenges for survival and reproduction. Early hominins likely faced these same challenges, so seeing how nonhuman primates overcome them in their environments today is useful for modeling the needs of early hominins and identifying what strategies they likely employed in order to meet them.

Evolution of the Human Diet

Paleoanthropology is a unique combination of biological anthropology and archaeology. The field is concerned with both the biological evolution and the culture of our ancestors, so researchers often work with both fossil records and archaeological records that date to before 10,000 years ago. Although some paleoanthropologists reconstruct our ancestry as far back as our origins in the Primate order, most work is done to address two central questions: What defines the hominin lineage? and What defines modern humans?

These two questions lie millions of years apart on the evolutionary timeline. To answer the first question, paleoanthropologists rely greatly on what we know of our living primate cousins because defining the hominin lineage requires an understanding of how early fossil hominins differed from other apes. About five to six million years ago, our lineage diverged from the branch that leads to present-day members of the genus *Pan*, which includes chimpanzees (*Pan troglodytes*) and bonobos (*Pan paniscus*). It is not a straight shot from the origins of the lineage to the evolution of our species, *Homo sapiens*. Thus, in trying to answer the second question related to our modernity, we must untangle a web of information for which the fossil record preserves only fragments.

Today we are the only surviving hominin species, but many hominins came before us. Two of the challenges for the field are determining how all the fossil species relate to one another and identifying which species was the direct ancestor of humans at any given time. At about 3 million years ago, we begin to find evidence of hominin material culture in the form of modified stone tools. This marks the beginning of the archaeological record. Tools are most commonly used for some purpose related to food.

A simple tool, even a stick that would not have preserved in the archeological record, would have enabled an individual to intensify their use of natural resources: all of a sudden a food that had been previously unavailable would have become accessible. Over the evolution of the hominin line, brain sizes got bigger and tools became more complex. These two changes increased the breadth of hominin diets even further.

For a variety of reasons, paleoanthropologists are interested in what foods hominins were eating and in what proportions. One reason is that a species' diet is interlinked with other aspects of behavior, such as sociality and geographic range, both of which are important for reconstructing past ways of life. Another reason for studying diet, and one that may be more relevant to this book, is that humans have invented many ways of feeding in extremely varied environments. These adaptations likely have an evolutionary history based on what foods are available, accessible, and desirable in a particular region. Most human evolution took place on the African continent, and edible insects are abundant in the tropics and subtropics. The following chapters make the case that insects were an important resource during this time in our evolution and that the practice has persisted to modern times, shaped by both environmental and cultural factors.

* * *

Before diving into human evolution, in chapter 2, I examine why we do not see edible insects in what is commonly referred to as western culture. Similar to how we understand India's sacred cow, insect avoidance cannot be fully understood from one perspective alone. I investigate the psychological concepts of fear and disgust as they relate to edible insects and the environmental and cultural history of Europe and North America. By integrating these concepts, I am able to more accurately portray why we are so strongly averse to insects.

Chapter 3 looks at ethnographic examples of the consumption of edible insects, focusing on populations of interest to human behavioral ecologists. In these populations, the environment plays a major role in food availability and ultimately in the capacity of a culture to survive and reproduce. Across cultures, women tend to forage and eat insects more than men. The question is whether this difference can be attributed to the nutritional benefits women receive from giving more attention to insects than to other resources. Evolutionary theory can help answer this question.

Chapter 4 examines how nutritional demands and evolution intertwine in the process of natural selection. Much of the discussion of evolution of the human diet centers on energy requirements because caloric demand increases as brain and body sizes increase (Leonard and Robertson 1997). However, nutrients play other roles in the body, such as building the structural components of the body and regulating physiological processes. The latter functions are especially important from the viewpoint of pregnancy and child-rearing. Reproductive demands differ between the sexes, a fact that adds to the discussion of why the different sexes tend to target different food items. Differential resource procurement patterns between men and women in foraging societies may have been shaped by natural selection to ensure that nutritional requirements were met.

In chapter 5, I review how different species across the Primate order consume insects. Although some species are insectivores, many primates use insects as a supplement to diets that otherwise are comprised of plant foods such as fruit and leaves. Patterns of insect eating in these non-insectivorous primates may reveal how this resource helps primates meet nutritional requirements. Interestingly, there is good evidence that female primates consume more insects than their male counterparts. Although this pattern is well recognized for tool-using chimpanzees and orangutans, similar evidence exists for other primates, such as capuchins and mangabeys. The widespread nature of this pattern suggests that it is due to different dietary requirements of the sexes. If human females and nonhuman primate females eat more insects than their male counterparts, then a case can be made that females of the last common ancestor and its female descendants on both the human lineage and chimpanzee lineage likely also ate more insects.

The goal of the book is to reconstruct the role of insects over the course of human evolution. I have combined behavioral accounts with fossil evidence to reconstruct past diets and determine the potential role edible insects played. Chapter 6, the first paleoanthropology chapter, focuses on the early hominins, especially those known as australopithecines. These small bipedal hominins are well represented in the fossil record from about four million to about two million years ago. It is with this group that we find the earliest artifacts of tool use. Among the tools australopithecines used were digging implements made of bone that preserved wear patterns consistent with digging into termite mounds. These tools provide direct evidence of insect foraging by human ancestors.

Chapter 7, another paleoanthropology chapter, shifts the focus to later in human evolution and members of the genus *Homo*. With the evolution of *Homo erectus* about two million years ago, morphology and behavior begin to resemble our own. Larger brains, taller stature, and more complex tools are hallmarks of *Homo erectus*. Edible insects would have been available in abundance to these hominins in Africa, but *Homo erectus* was the first hominin species to colonize outside Africa. Outside the tropics, the predictability and reliability of insects as a food source are greatly reduced, and the first occupants of regions like present-day Europe likely had little to no insects in their diets.

Reconstructing the behavior of past hominins has numerous limitations. Scientific discovery requires evidence-based research, but the paleoanthropological record preserves only fragments of past ways of life. However, the models I have created with data collected from extant populations establish hypotheses and predictions for this hominin behavior. This is the important first step in the scientific process. Chapter 8 investigates how future research can test these models and inform us about the ancient use of insects as food. The models I present were created from currently available data; more research that directly investigates the insect portion of extant diets and increasingly standardized data reporting will help us refine these models. I explore the numerous analytical methods that are available for the reconstruction of past diets and determine their potential for informing us about hominin use of edible insects.

Finally, in chapter 9, I address why reconstructions of hominin diets focus disproportionately on meat consumption and why this imbalance leads to underrepresentations of women in reconstructions of past behavior. Additionally, the bias toward meat consumption leads to portrayals of our ancestors as practically carnivorous in popular accounts. In recent years, such portrayals have been incorporated into the "paleo diet" fad. As a trendy weight loss program, the paleo diet emphasizes eating natural foods that would have been available to our "cavemen" ancestors. The emphasis on natural food is a direct response to our overindustrialized food systems, which produce widely available, inexpensive, unhealthy food options. Another major issue with our modern food system is that it is unsustainable: livestock cultivation is primarily responsible for wasted resources and greenhouse gas emissions. We should be looking to reduce our meat intake, not increase it. Edible insects provide an appealing option; raising them is efficient and sustainable and they provide the same nutritional benefits as traditionally raised livestock. Insects are a real paleo food, and the popularity of the paleo diet may offer the greatest potential for adopting insects as a part of our society's cuisine.

2

Understanding the Ick Factor

As a scholar studying the role of edible insects in the diets of our ancestors, at some point I began to feel like a hypocrite because I could list their numerous benefits yet I did not eat them myself. With some effort, I managed to overcome the psychological barriers, which was no small feat. I, too, am a product of my culture, and I had a strong revulsion to the idea of eating insects; but I found that knowing why I had this reaction helped me overcome it. In this chapter, I identify some of the issues that trigger negative reactions to eating insects. I also demonstrate that our aversions are not entirely biologically driven and that many people are rising to the challenge to overcome these deep-rooted biases.

Fear, Disgust, and the Concept of Bugs

In the search to understand why someone might be averse to edible insects, a good place to start is the evolutionary advantages of fear and disgust. These topics fall more in the field of evolutionary psychology than anthropology but are still relevant here. Evolutionary psychology attempts to identify behaviors that are universal to humans because they originate deep in our ancestral past. Some psychological processes, such as the basic emotions, may exist because they had an advantage related to selection in the evolutionary process. We share some of our emotions with other animals, which suggests that they originate in a deeper past. However, it can be difficult to identify unique human cognitive processes that have an evolutionary origin because culture has shaped much of what we do and how we think. Behaviors that may seem universal in psychological experiments administered to university students may not hold up to cross-cultural investigation. Additionally, some of the variation at the cultural level may be related to environments in which cultures developed that either promoted or discouraged psychological predispositions.

Some of the aversion to edible insects may come from fear. Although psychological literature on the fear of insects exists, a larger body of work exists for the fear of snakes, so that is a good place to start. Many people have a fear of snakes. Our nonhuman primate cousins also often display this fear, which is to be expected since snakes are one of their natural predators. The places where this fear appears in the Primate order can hint at its evolutionary origin. In fact, snake avoidance may be responsible for unique primate visual systems that include visual specialization for color and object form which are linked to brain size expansion in higher primates (Isbell 2006). Venomous snakes and small, fast-moving mammals, such as rodents and primates, appeared in the fossil record around the same time, about 60 million years ago. In areas such as Madagascar, where primates evolved without the pressures of highly venomous snakes, the primates have poorer central vision and visual discrimination abilities. In contrast, the primates of the division Catarrhini, which includes Old World monkeys and apes (including us), are the only group who have co-existed with venomous snakes without interruption. This is likely why the catarrhine brain evolved the most sophisticated visual centers found in the order Primates (Isbell 2006).

Despite these evolutionary forces, it is inaccurate to say that the fear of snakes is universal in people today because some people keep snakes for pets and enjoy handling them. It has been demonstrated in both human and nonhuman primates that a fear of snakes has a developmental learning component. In an experiment where infants were shown videos of snakes, the babies did not react differently to the snakes than to any other animal. In a follow-up experiment, babies prolonged their gaze on the snakes when the video was accompanied by audio of a frightened human voice, but not when the voice was happy. When babies looked at videos of animals that were not snakes, there was no difference in how long they gazed when happy or frightened voices were played. These results do not hold when photos are used instead of video, however, which suggests the trigger lies in the visual stimulus from the motion of the snake. The authors of this study concluded that humans have an innate predisposition to be afraid of snakes, but it also seems that this fear may not be actualized if the infants never experience the fear of other humans (DeLoache and LoBue 2009). Similarly, lab-reared monkeys do not initially demonstrate a fear of snakes, but they acquire it rapidly when they observe other monkeys showing a fear of snakes (Mineka and Cook 1988).

Although some snakes, like garter snakes, are essentially harmless, they move with the same slithering motion as their venomous counterparts and can still trigger fear. This broad aversion to snakes helps stop people from interacting with the wrong one. A similar argument could be made for insects, or better yet, for what many people refer to as bugs. In this context, bugs are a nontechnical group of invertebrates that often includes all land arthropods, including spiders and scorpions, and almost anything else that could be considered creepy or crawly. However, to entomologists, the term bug has a taxonomical meaning, referring to the order Hemiptera, which includes aphids, cicadas, and leafhoppers, to name a few. Because the word bug has accrued a broader, more casual meaning, entomologists refer to members of Hemiptera as "true bugs." In the nontechnical sense of the term, many bugs have harmful defenses such as stingers and venom. It is likely that a fear of bugs, whether warranted or not, comes from a similar developmental conditioning as that seen for snakes.

Fear of bugs is most commonly manifested as a fear of spiders (Gerdes, Uhl, and Alpers 2009). Like snakes, spiders are predators that sometimes use venom to subdue their prey, and the bite of some spiders can be deadly to humans. There is a lot of literature on spider phobias, but whether there is an innate pathway to this fear has not yet been determined (Rakison and Derringer 2008). Additionally, arachnophobia is less common in nonwestern cultures, and no studies show that nonhuman primates demonstrate a fear of spiders. It appears that a fear of spiders is more a product of one's culture than a mechanism of biological preparedness.

A reaction of repulsion in response to bugs may be better understood as disgust rather than fear (Looy, Dunkel, and Wood 2014). The disgust emotion promotes the avoidance of contamination and disease-carrying microbes. In a cross-cultural survey with respondents from 165 countries, respondents were asked to rate the level of disgust triggered by twenty photographs that appeared one by one on separate web pages. Among these twenty photos were seven pairs of images in which one depicted a disease-salient stimulus and another was matched to be as similar as possible but had no disease relevance (Curtis, Aunger, and Rabie 2004). More than 98 percent of people found the disease-relevant photo equally or more disgusting than the photo that was paired with it. For instance,

Figure 2.1. Although caterpillars may trigger disgust for some, they are commonly consumed in many cultures around the globe, including these mopane worms (*Gonimbrasia* spp.) collected in Limpopo province, South Africa. Photo by J. Lesnik.

a handkerchief with a blue stain was half as likely to trigger disgust as its red- and yellow-stained counterpart, which elicited thoughts of bodily fluids. Even the images of disease-salient invertebrates (a louse and some parasitic worms) were rated significantly more disgusting than their respective pairs (a wasp and some caterpillars), although study participants ranked the non-disease-salient invertebrates as more disgusting than the other images not associated with disease (figure 2.1).

The invertebrates that have no direct relation to disease may trigger a different kind of disgust. Psychologists categorize different forms of disgust in multiple ways. Two other forms besides contamination disgust are core disgust and animal reminder disgust (Rozin, Haidt, and McCauley 1999; Olatunji et al. 2008). Core disgust means disgust as it originally functioned, for example avoiding foods that are contaminated or otherwise offensive. The presence of insects on food is often a signal of spoilage, whether directly, as with maggots, or indirectly, as with cockroaches, flies, or other insects that can carry contaminants. In the United States and other countries where most people lack the ability to distinguish among the large array of invertebrates known as bugs, it is common to project these filth-related reactions onto the innocent, like caterpillars. The very common fear of spiders may in fact stem from a misplaced association of spiders with disease epidemics that developed during the Middle Ages in Europe (Davey 1994).

Animal reminder disgust includes anything that reminds us of our own animal nature (feces, sex, blood, etc.), but it also extends to the food we eat. Eating identifiable body parts or any reminder that meat products were once body parts can trigger a similar response. Most Americans are far removed from the meat they eat; they often buy it in plastic-wrapped packages after it has already been butchered elsewhere. Since they rarely encounter animal reminders in their food, even though they might eat meat every day, the thought of eating an insect with legs, eyes, and antennae can be a strong disgust trigger.

Our bodies have another system in place for avoiding harmful foods, and that is through our taste mechanism. As omnivores, we evolved to eat a large variety of foods, and taste buds enabled our ancestors to determine whether a food was worthwhile. If foods were sweet or had a desired amount of saltiness, then there may have been a benefit to our ancestors if they increased foraging efforts for those foods. Conversely, many toxic compounds taste bitter, signaling that they should be avoided. Even today, when we ingest bitter foods, oral receptors may trigger responses adapted to minimize poisoning, such as gagging or vomiting. The toxins that many insects produce as chemical defenses can trigger these taste reactions. But not all insects are poisonous or bitter, and anticipating bitterness is generally not what dissuades people from trying edible insects. In fact, most descriptions of edible insects liken them to appealing foods that people are familiar with, such as shellfish, pine nuts, sunflower seeds, mushrooms, and even bacon (Martin 2011).

The sensation of taste involves much more than activating taste buds. Texture, sounds, and other properties of foods are also important elements of insect consumption (Kouřimská and Adámková 2016). Slimy, gooey, and mushy foods have been correlated with adverse textural properties and disgust triggers (Breslin 2013). Raw insects would definitely have some of these textural properties, and many people fear this aspect of eating them. Even though insects are commonly consumed after they are roasted, cooked, or dried, which eliminates most of these properties, people still anticipate a negative experience (Bodenheimer 1951).

This anxiousness may better be defined as a neophobia, the fear of something new. The culture we are raised in creates an environment that teaches us what is safe to eat. Food neophobia is common in early childhood but usually decreases by adulthood as the result of positive food experiences. However, some people maintain high levels of food neophobia as adults. In a study that compared the reactions of people in Germany and China to a range of edible insect foods, the Chinese, who were more familiar with eating insects, had a more positive reaction to the foods, while Germans, who were less familiar with insect consumption, were less willing to eat the foods that included insects. However, individuals from both cultures who ranked high on the food neophobia scale demonstrated an equal unwillingness to try the edible insects (Hartmann et al. 2015).

The psychology of disgust is complicated, and it seems that edible insects can fall into a variety of categories of disgust triggers that can be picked up in early childhood. What is clear, though, is that there is nothing inherent about insects that would cause people to have adverse reactions to them, as evidenced by the billions of people who eat them today. Global patterns of insect consumption and avoidance can help us understand why these aversions developed.

Global Patterns of Insect Consumption

About half of the world's countries have cultures that include insects in their diets. We know of over 2,000 insect species that are consumed (Jongema 2017). The countries whose populations generally do not eat insects include the United States, Canada, and many members of the European Union. This pattern led many researchers to question why people living in western cultures do not favor eating bugs (map 2.1). In addition to associating insects with disease, it is commonly suggested that insects are seen as the pests of food crops. One suggestion is that this stigmatizes them as nuisances and leads people to avoid them as food (Harris 1985; Van Huis et al. 2013). If people avoid eating edible insects because they associate them with agricultural pests, it should be expected that countries that have high proportions of unfarmed land would eat more insects. If people do not eat insects because they see them as carriers of disease, then insect avoidance should be higher in high-density populations where disease vectors are more problematic. However, increased population density might suggest food insecurity, so perhaps people in those populations consume insects as a fallback food even though insects have the potential to transmit disease.

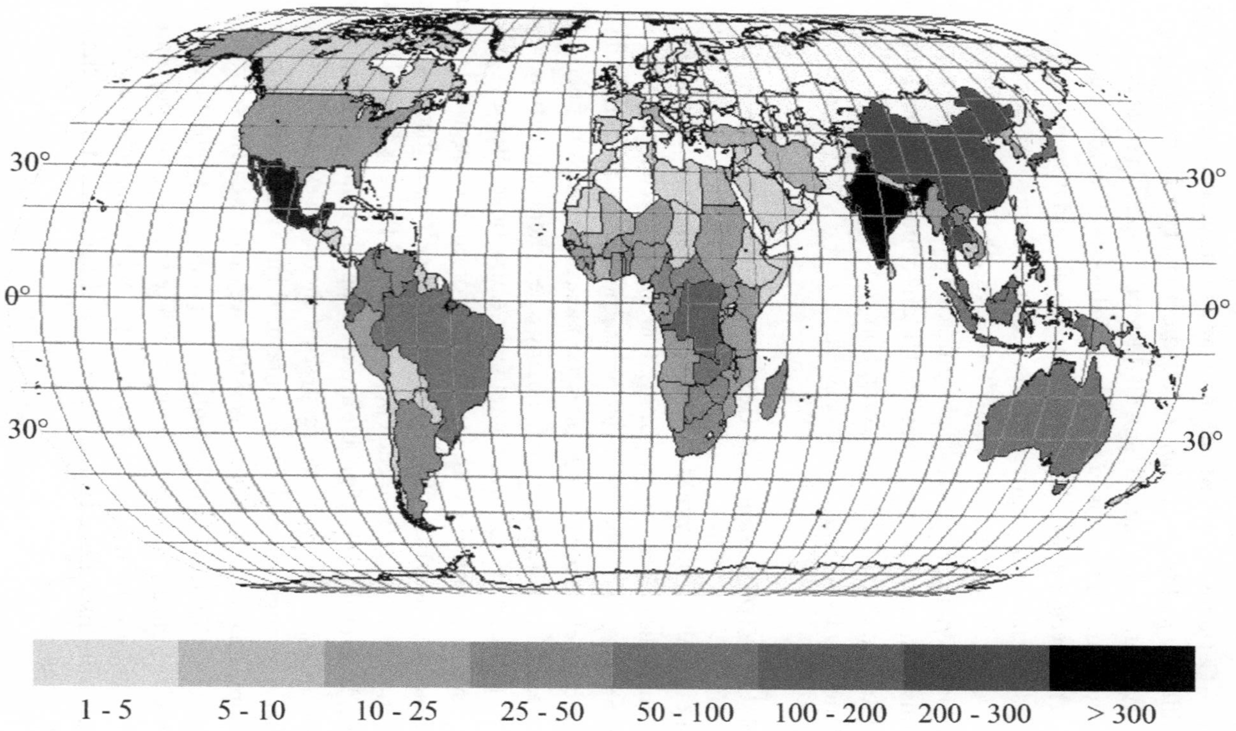

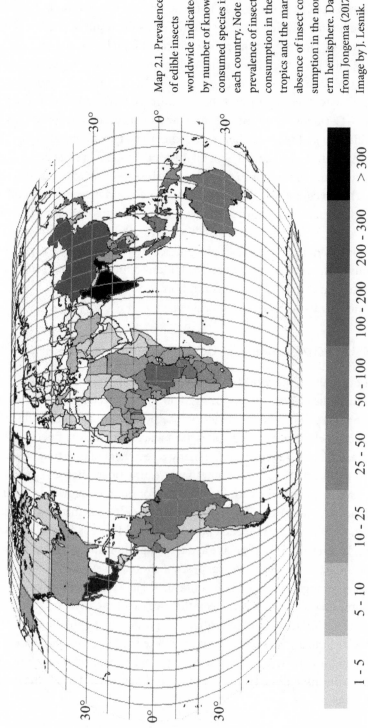

Map 2.1. Prevalence of edible insects worldwide indicated by number of known consumed species in each country. Note the prevalence of insect consumption in the tropics and the marked absence of insect consumption in the northern hemisphere. Data from Jongema (2017). Image by J. Lesnik.

When I tested these hypotheses, I found no correlation of edible insect consumption with population density or with the amount of farmable land (Lesnik 2017). The strongest predictor of insect consumption was latitude; the latitude of a country correctly predicts the practice of consuming insects 80 percent of the time. Insects are most commonly consumed in tropical regions and the likelihood that people will eat them is lower in countries farthest from the equator.

The data that I used to test these hypotheses came from nationwide statistics, and thus they do not address why any single culture within or across countries consumes or avoids insects. The data are only useful for identifying overarching patterns, and the only universal that exists is one based on latitude. This result suggests that people who live in latitudes not close to the equator may not eat insects for reasons attributed to their non-tropical environments, but this alone cannot explain why people living in these latitudes actively avoid insects.

From Absent to Averse: Western Culture and Edible Insects

It is commonly stated that insects are not consumed in western culture. Etymologically, the word "western" refers to nations across Europe, but the term's meaning has been broadened to include countries with a history of European colonization and continued European migrations. However, the idea of a singular culture across these countries is problematic because not every nation with a history of European conquests shares the same economic, social, or spiritual leanings. Even more controversial is that the concept of western is argued to maintain the colonial perspective of "the west versus the rest," which devalues cultural diversity and ethnic autonomy. Both the United States and Canada are typically considered western, and like European countries, they have little to no insects in their diets. In Australia, over fifty species of insects are consumed, but it is the indigenous cultures that live mostly in remote central areas that eat them and insect consumers are essentially absent in metropolitan areas that share European economic and political points of view. This pattern also holds for countries such as Mexico and South Africa, where European influence has transformed modern cities. In such cities, edible insects are not consumed by the urban population but instead are tourist fare that appropriates and romanticizes traditional ways of life present in rural areas

Figure 2.2. Edible insects are sold in rural markets in South Africa but are difficult to find in large cities such as Johannesburg. In Thohoyandou, South Africa, mopane worms (*Gonimbrasia* spp.) and termites (*Macrotermes* spp.) are sold alongside wild-foraged plant foods in the marketplace. Photo by J. Lesnik.

(Goertzen 2010) (figure 2.2). It is clear that European influence reduces the consumption of insects. What needs to be addressed is why.

Insects are consumed the most in tropical regions, and the frequency of their consumption decreases as distance from the equator increases. This pattern correlates with a well-documented ecological phenomenon known as the latitudinal gradient of diversity (Gaston 2007). Biodiversity is greatest in the tropics, or equatorial regions, and species richness reduces gradually toward each pole. There are several reasons for this pattern. First, solar radiation is greatest in the tropics, providing plants with the resources they need for photosynthesis. This results in more plant life that can sustain other organisms higher up on the food chain. Second, the tropics have undergone less climatic variability over time than the habitats farther from the equator. The climate has fluctuated during the history of the earth, but even during the most dramatic changes elsewhere on the planet, the tropics have not deviated substantially from the long-term norm. In areas far from the equator, such as northern present-day Europe, climatic history is characterized by ice ages that vastly changed the ecology and significantly reduced biodiversity (Hewitt 1999). Even in the warmest periods, species richness cannot rebound to match that of the relatively unaffected tropics. The food choices available to people living in what is now Europe were affected by this climatic history. Clearly this does not mean that there are no insects in Europe, but there has been a significant reduction in species richness there compared to regions closer to the equator. In colder periods, there would have been even fewer species than there are now. Where there are fewer insects, the principles of probability suggest that fewer edible insects would be available, so the likelihood of insects being adopted as food in this region would be lower than in other parts of the world.

There are some examples of insect consumption in European history, but we have to go quite far back to find them. The ancient Greeks and Romans recorded that they ate locusts, grasshoppers, cicadas, beetles, and probably other insects. The most cited examples come from Aristotle and Pliny the Elder. Aristotle, a Greek philosopher in the fourth century BCE, wrote about the natural history of the cicada in in *Historia Animālium* (DeFoliart 2002). Besides being fascinated with their natural life cycle, Aristotle also found them delicious and noted the most efficient way to harvest them. Pliny the Elder, or Gaius Plinius Secundus, was a Roman naturalist during the first century CE. He described how the larvae of the *Cossus*, which has been determined to be the beetle *Cerambyx heros*, was highly regarded as food, especially when they were reared on flour and table wine (DeFoliart 2002).

It seems that by the onset of the Age of Exploration in the fifteenth century, these practices were essentially lost. A common presumption is that the spread of Christianity across Europe brought opposition to insect consumption, perhaps borrowing from Judaism the concept that insects were not kosher. However, even though the book of Leviticus in the Old Testament states that "all winged insects that go on all fours are detestable to you," it continued,

> Yet among the winged insects that go on all fours you may eat those that have jointed legs above their feet, with which to hop on the ground. Of them you may eat: the locust of any kind, the bald locust of any kind, the cricket of any kind, and the grasshopper of any kind. But all other winged insects that have four feet are detestable to you. (Leviticus 11:20–23)

The restrictions in the Old Testament could explain why Pliny's *Cossus* larvae and Aristotle's cicadas fell out of favor (larvae have zero legs and cicadas have six legs) but locusts and grasshoppers were still seen as consumable. Not only are they explicitly listed as acceptable food in the Old Testament, but they also appear in the New Testament, where the Gospel of Matthew recounts that John the Baptist ate locusts and wild honey. The spread of Christianity is not a sufficient explanation for the loss of edible insects in the first millennium CE.

Climatic history of the region may provide another clue for why insect consumption declined historically. Prior to our current warming trend, there was a period of global cooling known as the Little Ice Age that began in the thirteenth century and continued through the nineteenth century (Mann 2002). Although this was not technically an ice age, winters were much colder during this period, and harsh storms were common year-round. Species richness at this time in Europe must have been lower than it was previously and potentially lower than it is today. It is possible that the decline in insect consumption in Europe is related to the fact that the environment was no longer suitable for supplying the resource.

It was from this relatively cold environment, devoid of edible insect culture, that explorers such as Columbus launched their expeditions (Carew 1988a, 1988b). For those who settled the regions that Europeans colonized, food was an important part of identity. For instance, in the Americas, food was central to defining differences between Europeans and the indigenous inhabitants (Earle 2010). Spanish settlers were consistently preoccupied with how to access European foodstuffs and feared the deleterious effects of native foods. They expressed particular scorn for insect consumption (Earle 2010). The concept of eating insects was foreign to European settlers, and the practice was an easily observable factor that distinguished, in their minds, between "civilized" settlers and "savage" natives. Many indigenous cultures lost the practice of eating insects as they adopted the new European ways. Those who maintained this aspect of their foodways lived outside the centralized cities. This pattern still exists today; people who live in metropolitan areas typically equate insect consumption with poorer peoples who live in rural areas. Globalization has exported these views to people living in nonwestern countries, and younger generations are turning away from traditional insect foods. Unfortunately, many of these people live in areas of food insecurity, and in the absence of an alternative to insects in their diets, malnutrition is a serious threat (DeFoliart 1999). Thus, it is possible that a change in the opinions of westerners regarding edible insects may have greater benefits than reducing the environmental footprint of livestock.

Overcoming the Bias

Many people today are working to promote the consumption of edible insects. These people have already overcome deep-rooted biases. Increasing amounts of research is being dedicated to determining the most effective ways to convince more people to give insects a try. A study from Belgium facilitated a survey to determine which western consumers will be the most likely to add insects to their diets (Verbeke 2015). Researchers randomly selected 368 consumers of meat to fill out an online questionnaire. They told the participants that insects "are a good source of high-value proteins, their production requires little space, their feed conversion is efficient, and therefore the eating of insects provides benefits in terms of sustainability" (149), then they asked them to fill out the questionnaire. In addition to answering whether they would be likely to adopt insects into their diets, participants also provided information about what factors affect their food-buying decisions in general.

The researchers found that neophobia was the primary factor that deterred people from consuming insects but that younger males tended to be the most adventurous eaters. Males were more than twice as likely as females to adopt insects as a meat substitute. A ten-year increase in age was associated with a 27 percent decrease in the probability that a participant was ready to adopt insects as a food. Age and gender were not the only factors, however. Consumers who indicated that they intended to reduce their meat intake were 4.5 times more likely to be ready to adopt insects as a substitute for meat than consumers who did not plan to reduce their meat consumption. Individuals who said that they consider the environmental impacts of foods when making decisions about what to buy were also more likely to adopt edible insects. Thus, the most likely people to adopt insect consumption are younger males with a weak attachment to meat who are more open to trying novel foods and are interested in the environmental impact of their food choices.

The study identified two other important points that are relevant to the advancement of insects as food. Participants who were already familiar with the idea of eating insects were 2.6 times more likely to be ready to adopt insects than those who said they hardly knew anything about it. Additionally, there was a significant relationship between participants not prioritizing convenience when making their food choice and the likelihood that the participant would accept insects as food, which suggests that if insects were easily available as a snack or an ingredient in a convenience food, people would be more likely to try them (Verbeke 2015). Another study from Belgium found that western consumers are ready to buy insects and cook them at home if they are able to season and prepare them in ways they are already familiar with (Caparros Megido et al. 2014). These findings are encouraging because they suggest that the efforts to promote insect consumption may indeed have a positive impact and that startup companies that are incorporating insects into foods such as protein bars, crackers, and cookies are on the right track. What can also be gleaned from this chapter is that efforts to overcome bias against insect foods should take into account the fact that fear and disgust have developmental learning components. It will be important to remove colonial attitudes from dialogue on edible insects and provide children positive experiences with insect foods if we are to see long-term, generational changes in attitudes about eating insects.

* * *

In most countries outside Europe and North America, some amount of insect consumption is present. A glance at European history suggests that it was once present there also. It appears that environmental factors may account for a reduced dependence on the resource but that cultural factors associated with European colonization produced the stigma and disgust that westerners feel today. In other parts of the world, especially in the tropics, edible insects carry less of these negative connotations; they are perceived as food that is just the same as any other food. Insects are consumed in a wide range of societies, from some of the most powerful industrialized nations to the few remaining populations of hunter-gatherers living in remote areas.

3

Ethnographic Examples of Insect Foraging

Cricket farms and other facilities that rear insects for human consumption feed a limited number of people today. The vast majority of insects consumed around the world are collected through wild harvesting. I had the privilege of spending time with South African women from the township of Thohoyandou in the northern province of Limpopo who harvest soldier termites year-round and sell them in the local marketplace. These women regularly return to termite mounds they know are productive, often traveling by car, bringing buckets and tools such as garden hoes. They use the hoes to break open the mound, then they drop a broom constructed of nearby grasses and leaf strips into the opening. Termite soldiers attack the tool, gripping with their pincers and not letting go. It did not take long for me to learn that a bite from one of these soldiers cuts remarkably deep. However, while the termites are gripping the tools, the women are able to safely transfer them from the broom and into a bucket by quickly running a closed hand down the length of the broom. I, however, was not brave enough to try this technique without gloves. After a few hours of collecting, we returned home with our spoils: a bucket full of termites that were unable to escape up the slick sides of the plastic contianer. The women rinsed the termites with water, boiled them with some salt, placed them in a basket to dry, and took them to market (figure 3.1). These fresh-boiled termites were my favorite of any insect I have tried; they remind me a lot of popcorn. The sale of these treats provides an important supplementary income for these women (Netshifhefhe, Kunjeku, and Duncan 2018).

While it is important to understand the role of edible insects in local diets and economies like those of Thohoyandou so we can determine

Figure 3.1. A woman preparing termites (*Macrotermes* spp.) for sale in the marketplace in Thohoyandou, South Africa. Photo by J. Lesnik.

how best to promote their increased use to help address present-day food insecurity, this information is less useful for reconstructing the evolutionary origins of human insect-eating practices. Anthropologists who aim to understand how society functioned before large-scale civilizations developed rely on models that were created using the records of the few remaining hunter-gatherer populations. Hunter-gatherers, also called foragers, are people who live in a subsistence-level economy in which natural resources provide for all basic needs and there is no system of monetary exchange. In this chapter, I review selected anthropological literature on the role of edible insects in subsistence foraging. The published works included here describe which type of insects were foraged, by whom, for whom, and how often throughout the year this practice was done.

Cultural Correlates of Hunter-Gatherer Populations

Before the advent of plant and animal domestication about 10,000 years ago, people everywhere used only natural resources to survive. Although most foraging societies today have incorporated some aspects of modern life, such as domesticated foods and industrial products, understanding the natural environment and effectively extracting resources remains critical to survival. In the lush tropics, most of the diet of foraging societies consists of numerous and varied edible plants. In the Arctic, where there is less vegetation, most of the diet of hunter-gatherers comes from hunted or fished animal foods. Even though foraging peoples of different cultures eat different foods, they all live off the land, and thus there are many similarities in their lifestyles. These shared characteristics are known as the correlates of foraging (Kottak 2006).

People who subsist entirely on foraged foods live in small, band-level societies. A band is the most basic social unit that includes less than 100 people, all of whom are related by blood, marriage, or other kinship, who live communally for at least part of the year. Often seasonal limitations cause bands to split into smaller family-based groups. This enables people to spread out and exploit smaller patches of resources across diverse habitats. These smaller groups then reunite when seasonal change brings abundance, when cooperation is mandatory for success, or for ceremonies to celebrate life events or rites of passage.

Another typical characteristic of a foraging community is mobility. People move across the land not only because of the band's nomadic foraging practices or because of fission and fusion within bands but also because individuals change band membership. Marriage practices in these societies are exogamous, meaning that two wedded individuals must come from two different bands. Biologically, this helps prevent inbreeding since members of small bands tend to be closely related, but culturally, this gives the married couple kinship with at least two bands. This enables them and their children to move between the groups and connects multiple bands over a larger network, both of which can be critical to succeeding in uncertain environments.

Mobile foraging peoples do not have the ability to store resources in any large amount. Drying, fermentation, or other natural processes can preserve abundant foods, but the natural environment rarely produces any resource in abundance that requires storage in a central location. Because resources are not centralized, power among band members does not become centralized. In general, band-level societies are egalitarian; contrasts in prestige based on age and gender are small. Elders are revered for the knowledge they possess, as they have seen how their people navigated environmental fluctuations in the past. This knowledge can be critical to present-day success. Division of labor by gender is also a common practice of foraging groups. Men and women target different food resources and both bring back what they find to be shared. How the resources are divided and how much each gender contributes depends on the environment and the specific culture. Most often, men go after high-risk resources such as large game and women focus on collecting smaller, more predictably available plant and animal resources.

Since edible insects fall into the category of small and predictable foods, I hypothesized that women are the ones who acquire this food for their communities. In order to investigate this hypothesis further, I turned to the ethnographic record of foraging peoples to see how they used insects in their diets. However, this methodology has notable limitations.

Limitations of Available Ethnographic Data

An abundance of literature exists that documents insect foraging in one form or another. This includes letters written by explorers and missionaries and research conducted by anthropologists. Though the sources are numerous, very few provide detailed accounts of how foraging peoples collected, prepared, consumed, or generally valued insects. Yet this information is necessary for generating a model of insect-foraging correlates. It is to be expected that non-anthropologists who recorded their observations did not follow the best practices of ethnographic research, but even in anthropology, detailed accounts of edible insects are few.

As the modern world has expanded, the land hunter-gatherer populations used became cultivated with agricultural products and livestock. This process eliminated many critical natural resources. Even the few populations who have maintained their traditional lifestyles are not untouched by the modern world; they live in the political boundaries of nation-states, they often depend on some form of government to protect their rights to land use, and they have contact with outsiders who have introduced commodity foods and other goods. The opportunities to examine insect consumption in traditional hunter-gatherer diets have been dwindling consistently since anthropology became a field of study.

When anthropologists started to regularly conduct organized ethnographic research about other cultures in the late nineteenth and early twentieth centuries, the people they were most interested in studying had already been conquered and colonized, often by the anthropologist's nation of origin. For example, by the year 1950, anthropologists had established twenty-five ethnographic sites for studying indigenous populations in North America (Kottak 2006), but many of the peoples who lived on those sites had been removed from their native lands, by treaties, by violence, or by forceful ejection. Although edible insects were included in the diets of many indigenous peoples in North America, we know very little about how they used them. The richest environment for edible insects in the present-day United States is in the semitropical southeast, but the Indian Removal Act of 1830 forcibly relocated the major tribes of this region—Chickasaw, Choctaw, Creek, Seminole, and Cherokee—to the open plains west of the Mississippi River. The combination of the new environment and the scorn white people expressed about the traditional foods of native peoples undoubtedly caused changes in the diets of these groups, including a decline in, if not the total elimination of, insect consumption. The Cherokee of North Carolina, who were able to remain on a small trust of land the tribe purchased, have traditionally incorporated insects such as grasshoppers and wasp larvae into their diets, but very little literature exists that details this practice (Carr 1951; Chiltoskey 1975). Although there is a long history of ethnographic research of the North Carolina Cherokee, this food source is rarely discussed. Most likely, the lack of information about edible insects for indigenous groups of North America is because of disinterest on the part of the researchers, who by their own standards do not view the food source as valuable, and underreporting by community informants, since they are unlikely to describe in detail customs that elicit judgment from outsiders.

The most detailed accounts of indigenous use of edible insects in North America come from the Great Basin region. Unlike the experience of the indigenous peoples of the Southeast, outsiders did not occupy the lands of the Great Basin peoples until relatively late in U.S. history, when the Mormons arrived in 1847 (Smaby 1975). The fate of the foraging cultures there is the same as it was everywhere else; ultimately, they were relocated to reservations, where small land areas forced them to plant crops and raise livestock. The Great Basin peoples, however, most notably the Northern Paiute, were resistant to cultural assimilation. The Northern Paiute's Ghost Dance, a ceremony of renewal and rehabilitation, was a symbol of resistance, and many indigenous peoples of the Great Basin and beyond adopted it as their religion (Kehoe 2006). This dedication to maintaining cultural identity is likely responsible for the persistence of insects in the diets of these groups (Blake and Wagner 1987). Additionally, the Great Basin has a unique environment where seasonal swarms of grasshoppers drown in the area's saline lakes, creating an abundance of valuable food for foraging peoples living in a desert environment. Although these are not the only insects consumed in the region, this particular event draws large numbers of people to the lakes to collect the grasshoppers that wash ashore.

Over the years, anthropological research standards for evaluating other cultures have improved. Anthropologists have worked to eliminate ethnocentrisms, or projections of a researcher's cultural values onto the culture being observed. However, during the time it took anthropology to develop into a rigorous academic discipline, more hunter-gatherer groups lost their autonomy. Fortunately for this project, hunter-gatherers are not the only subsistence groups that provide data for modeling the benefits of insect foraging. Another relevant subsistence category is horticulture. Horticulturalists rely heavily on foraged foods, but they also maintain some cultivated crops. These peoples use non-intensive methods with their crops: they do not use irrigation or livestock, they maintain the fertility of the land by rotating plots, and they rely entirely on the environment for rain to water their crops. Horticulturalists' lives are directed by the natural environment almost as much as those of hunter-gatherers. Wet, humid climates, like the Amazon Basin, are the most favorable environments for this form of subsistence. Although tending crops requires horticulturalists to be less mobile than hunter-gatherers, they relocate their villages every five to twenty-five years to more productive lands to give previously cultivated fields time to recover. The productivity of this system often does not exceed that of hunter-gatherers, so population sizes remain small. However, when areas are especially productive, populations can grow to a few hundred people and more complex sociopolitical organizations develop (Kottak 2006). Some of the most detailed anthropological literature on the use of edible insects comes from studies of horticultural societies (Paoletti, Buscardo, and Dufour 2000) (map 3.1).

Below are detailed accounts of how different foraging societies incorporate insects into their diets. Collectively, these studies provide much ethnographic data that are useful for identifying correlates of insect foraging. However, because of the nature of studying small-scale indigenous populations, the data have limitations because the sample is scattered across continents.

Ethnographic Accounts of Hunter-Gatherers

The San of Southern Africa

The San is a general name for hunter-gatherer populations across southern Africa. The nutritional contributions of the gathered foods of the !Kung San of the Kalahari Desert have been studied extensively. San women have notably low fertility with long intervals between births and anthropologists have been studying their diets and lifestyles in order to understand this pattern. Researchers have found that the foods available to the !Kung are numerous and have high nutritional value, including the widely available and easily preserved mongongo nut. Although San men also collect resources while out on hunting trips, the women are responsible for much of the gathering. The energy expenditure the San women's workload requires plus the additional energy expense associated with traveling to visit neighboring kin are responsible for the lower birth rates, not the quality or abundance of food. Overall, the San are healthy and have the potential to live long lives (Lee 1979; Truswell 1977; Bentley 1985; Bogin 2011).

As Nonaka (1996) details, the San use some insect species year-round

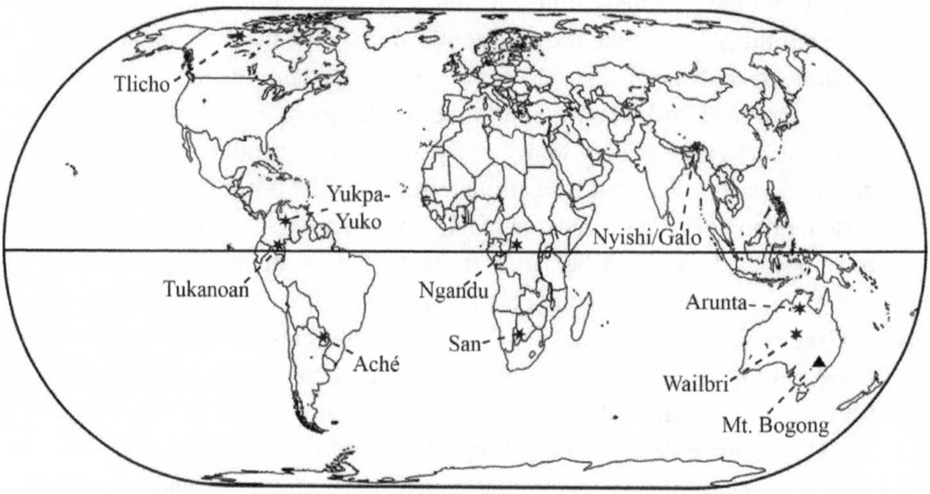

Map 3.1. Map of insect-foraging population locales featured in this chapter. Map by J. Lesnik.

but also take advantage of seasonal resources. For instance, during the breeding season of termites, massive swarms of flying insects known as alates form that are looking to establish new colonies. During the swarming nuptials of the harvester termite (*Hodotermes mossambicus*), women chase the swarms to identify their underground nests. Once a nest is located, the women enlarge the openings of the nest and fill them with grass to prevent the termites from escaping. Then they gather the entire swarm, along with the grass, and take it home, where they roast the termites on hot ash and sand. Other termites with nests above ground, known as mounds, are easier to spot while the women are foraging. At any time of year, women will dig out a mound with a stick, pick out the nymphs, and eat them raw. If the mound yields a large quantity, the women will stop gathering plants and sit down to eat all day.

Other termites create subterranean nests in the San camp area, but because of their inconspicuous location, they are difficult to find except during the nuptial flights of the alates. Children harvest these termites when they can find the nest; they surround the exit holes and wait for the termites to appear. They pinch the termites as they emerge, remove the wings, and eat them raw. Nonaka notes that these termites are snacks for the children; adults do not eat them because they are available only in small quantities (Nonaka 1996).

Another seasonally available insect is the caterpillar of the hawk moth (*Herse convolvuli*), an especially important resource for the San. During an outbreak of these caterpillars, women set up makeshift camps by their trees, where they stay for several days to collect as many caterpillars as they can. They roast and dry the caterpillars, then store them for months. The San enjoy the taste of these insects; they liken it to the taste of large game and use the caterpillars to add flavor to their meals (Nonaka 1996, 33).

Women also regularly gather grasshoppers, beetles, and ants. They find grasshoppers (*Cyrtacanthacris tatarica* and *Lamarckiana cucullata*) clinging to trees and huts in the mornings and late afternoons during the rainy season, when they are easily collected by hand. The women remove the legs and roast the bodies over hot ash and sand. After the grasshoppers are cooked, the women remove the heads and the connected entrails; the remaining body is what is eaten. By 1996, when Nonaka published his work, grasshoppers were already falling out of favor. Some of the elders still eat them, but most people reported that they did not like their grassy odor.

Buprestid beetles (*Sternocea orrisa*) and ants (*Camponotus* sp.) are more readily available throughout the year. Women collect the beetles in whatever numbers they find them, great or few. They either eat the beetles immediately (after removing their heads) or they roast them and ground them into a paste. They also often collect ants in small handfuls after disturbing their nests. They bring them back to camp and mix them into another dish such as a salad.

Nonaka provided an account of an experiment with another potential insect food. One old man said that his father and some other men tried eating the armored ground cricket (*Acanthoplus*). When they removed the thorns from the crickets' backs, juices oozed from the openings. The men assumed that these were the fatty oils they had seen in other insects, but after boiling them in their juices and consuming them, some of the men experienced significant digestive upset. The armored ground cricket has since been avoided. Nonaka noted three interesting observations about this story. First, the experimentation took place during a bountiful season, not one of need. These crickets were not being looked at as a fallback food but instead were being investigated as a possible treat. Second, it was the men who tried the new food. It was usually the women who gathered the insect foods, but because this was an experiment, it was the men who thought to try it since they were the ones to take more risks. Finally, the juices that came from the cricket seemed to be fatty, which was highly valued. The insects were boiled instead of roasted in order to use this ingredient, but the juices were likely part of a toxic defense mechanism and the men became ill as a result.

One final note about the Nonaka article is that he noted in the abstract that "even though insects are not an important subsistence resource, the San have an extensive knowledge and make good use of insects" (1996, 29). However, it seems that insects are quite important, since women will stop all day to forage for termites or will make an auxiliary camp for a week during a caterpillar outbreak. Because of the unpredictability of these episodes and the irregularity of these insects as food, they may not have seemed like an important component of the diet to Nonaka. My interpretation is that resources must be highly valued if their availability interrupts the regular routine. Additional evidence for the value of insects to the San comes from the pervasiveness of termite nest depictions in San rock art. San members have told ethnographers that these images relate to the San views of the spirit world, suggesting that termites evoked creative and transformative power (Mguni 2006). Seasonal swarms were often depicted in these images, suggesting that this time of year was worth celebrating.

The Aché of Eastern Paraguay

The Northern Aché were a hunter-gatherer population that lived isolated in the forests of eastern Paraguay until the 1970s when they were forcibly moved to reservations. The Aché are one of the most-studied groups of modern hunter-gatherers. Systematic studies have been done to determine how Aché men and women spend time foraging and conducting other related activities. A study of sex differences in food acquisition claimed that men produced more food than women because the women's participation in food gathering was limited by their child care responsibilities (Hurtado et al. 1985). However, according to the authors, "women walk several hours a day primarily to carry the game that men killed so that men can cover more territory while hunting" (4), yet they did not factor these hours into the women's activity budgets. Instead, they focused on how much time the women spent resting. Aché women could not have maintained their relatively high fertility rates (Hill and Hurtado 1996) if they did not make it a priority to take breaks between helping the men carry game and gathering their own foraged foods.

A detailed account of edible insects does not exist for the Aché as it does the San, but insects are included in descriptions of their general foraging patterns. The study investigating sex differences (Hurtado et al. 1985) reported that Aché women spent about an hour a day foraging for plant foods, while they spent an estimated average of fifteen minutes a day looking for various beetle larvae. Outside of this allotted time, they would collect these larvae whenever they encountered them opportunistically. When they found larvae, they almost always ate some immediately, then took some back to camp to boil in palm broth for the family. Thus, women consumed more of these insects than men. On occasion, men and women would forage together, and on one such documented instance a husband and wife spent over an hour collecting larvae from logs, resulting in 1.26 kg of larvae that they took back to camp (Hawkes, Hill, and O'Connell 1982). Although larvae of beetles such as *Ryhncophorus palmarum* (palm weevil) are major contributors to the diet year-round, in the 1980s, the Aché consumed twelve types of insects throughout the year, especially in the warm season, when insects are more active (Hawkes, Hill, and O'Connell 1982; Hill et al. 1984; Hurtado et al. 1985).

Australian Aboriginal Tribes

It is well documented that various aboriginal tribes of Australia eat insects. Unfortunately, no thorough ethnography exists that focuses on consumption of edible insects. Instead, we have some less detailed accounts from broader food surveys and some general overviews of aboriginal diets that lump all groups together (Cherry 1991; O'Dea et al. 1991).

The Arunta, Northern Territory

The Arunta, also known as the Arrernte, were traditionally foragers in the Northern Territory of Australia, but they have become increasingly sedentary since settlers moved onto their lands in the 1860s. Because they were desert dwellers, animal foods were an important part of their diet. Men hunted kangaroos and emus, and women, accompanied by their children, used digging sticks to search for small burrowing fauna such as honey ants, termites, and grubs. Social insects such as ants, termites, and bees (for their honey) were an important year-round resource in the dry environment. Nonsocial insects such as the witchetty grub (for the larvae of moths in the Cossidae family) were available only during short periods after heavy rains but were highly valued. The vital importance of these insects is reflected in the presence of gestures in the language; people used a specific gesture to indicate the discovery of witchetty grubs and another gesture to indicate the discovery of honey ants (Bodenheimer 1951).

The Wailbri, Northern Territory

The Wailbri, also known as Warlpiri, are another foraging population of central Australia. In 1947, when Sweeney wrote of the food supplies available to this desert tribe, there were only 600 members and about half of them had already left for places where they could more easily obtain food, such as reserves or pastoral stations. The Wailbri men and women foraged for the same insect foods the Arrernte used. Sweeney provided detailed information about how they foraged for honeypot ants. The worker caste of these ants feeds on nectar from flowers and stores this honey-like resource in their enlarged abdomens, from which other ants in the colony can then feed. The Wailbri women would dig into the ant colonies to obtain these engorged workers. Sweeney described how people would suck the globule of honey from the ant, although there are also accounts of the ants being consumed whole across Australia (Cherry 1991).

Bogong Moth Harvest, New South Wales

Hundreds of people from several different aboriginal tribes would come to harvest the Bogong moth (*Agrotis infusa*) in the Bogong Mountains of New South Wales. The Bogong moth occupies the crevices and caves of mountains, where it goes into a state of dormancy, known as aestivation. This adaptation is similar to hibernation in cold climates, where metabolic rate and activity levels drop in order to enable an animal to wait out the extreme temperatures, but with aestivation, it is the hot season that initiates the sleep. Layers of these moths would cover the caves, their state of torpor making it easy to collect by the people who came to feast. Cooking them in sand and hot ash burned off their wings and legs, leaving the cooked bodies to be consumed or made into paste or cakes (Flood 1980).

The Yiatmathong tribe historically controlled the valleys that provide access to Mount Bogong. Other tribes participating in the feast would need to gain permission to access their land and would gather to await passage. Kneebone (1991) describes the gathering as follows:

> Many ceremonies would have taken place, initiations, marriages, trading, settling of disputes, renewing alliances and friendships. When this was over one last ceremony had to take place before the trek to the alps and the Bogong Moth Feast would begin. This ceremony was receiving of permission to travel over someone else's territory. (3)

Gathering these moths provided both a seasonally available nutritional resource and the occasion for an important networking event for neighboring tribes.

Nyishi and Galo Tribes, Northeast India

Arunachal Pradesh is the largest state in northeast India. The mountains and valleys of this region create complex tropical and subtropical environments and great biodiversity. Traditionally, communities such as the Nyishi and Galo tribes lived exclusively from foraged resources. They maintain many of these practices today, although they have added nonintensive agriculture to their economies. Chakravorty and colleagues (2011) studied the extensive use of insects in the diets of these two communities. Although they do not report on the methods used to collect any of the insects, I include the study here because of the diversity of edible insects it describes.

Chakravorty and colleagues found that the Nyishi and Galo people consumed at least eighty-one species of local insects belonging to five orders: Odonata, Orthoptera, Hemiptera, Hymenoptera, and Coleoptera. There is a marked absence of termites (Isoptera) and moths, butterflies, or caterpillars (Lepidoptera) in the diets of these two groups. Certain insects, such as weaver ants, are available all year long, but most are seasonal. The seasonal availability of the different insects overlaps considerably, and multiple edible insects are available at any time.

Members of the Nyishi and Galo tribes consume twelve of the edible insect species for medicinal purposes. Bee and wasp species were most widely used for therapeutic purposes. Wasps are chewed (but not swallowed) to treat coughs, colds, and stomach disorders. Ants, which contain formic acid, are used in treatments for scabies, malaria, toothaches, blood pressure anomalies, and other ailments. Although healers from a number of cultures around the world use edible insects for medicinal purposes (Srivastava, Babu, and Pandey 2009), the efficacy of these products is poorly understood. It is suspected that the natural defenses of the insects can transfer their antifungal, antiviral, and antibacterial properties to people, but there is a need for rigorous scientific investigation to accompany these ethnographic accounts.

Arctic Foragers: The Tlicho, Northwest Territories, Canada

Although edible insects are largely a tropical resource, they are not absent from the diets of foragers at more northern latitudes. In the early twentieth century, caribou hunters such as the Tlicho (Tłı̨chǫ) of the Northwest Territories were known to eat warble fly larvae (*Oedemagena* or *Hypoderma*), a by-product of their hunting (Felt 1918). Hunters often found the larval form of this parasitic fly species in abundance when they were butchering caribou because the painful infection the parasite caused would make their hosts weak and thus easy prey for hunters. The Tlicho valued these larvae for their taste and would often leave them in place to develop further before eating them raw.

Additional Ethnographies from Horticulturalists

The Tukanoan of Amazonia

The Tukanoan (or Tucano) are indigenous people in the Amazon region of Colombia who practice horticulture, primarily the cultivation of cassava, but still extensively forage for naturally available foods. Acquiring protein can be challenging for them now that they have begun growing agricultural crops, and how they meet these requirements has been of interest to anthropologists. The dense rainforest provides a great biodiversity of fauna, including large quantities of insect prey. In one of the most detailed accounts of edible insect use, Dufour (1987) studied the Tukanoan as a case study of using insects as food.

Figure 3.2. The Tukanoan people of the Amazon region of Colombia collect palm weevils (*Rynchophorus* spp.) in larval form from fallen trees. Photo by Paul N. Patmore.

In the 1980s, the Tukanoan used over twenty species of insects as food resources, but the most important to their diet were the ones that aggregated in large numbers (social insects such as termites and ants) and the larval forms of insects that reproduce in large numbers (such as caterpillars and beetle larvae). Dufour found that insects were most important in the diet of the Tukanoan during the latter part of the rainy season in May and June. During these months, insect consumption compensated for the limited availability of fish or game; they provided up to 12 percent of the animal protein in men's diets and 26 percent in women's diets. Most of the protein came from ants and termites, which were almost exclusively foraged by the women.

The soldier caste of both ants and termites (genera *Atta* and *Syntermes*, respectively) were consumed year-round; women would fashion a tool out of a palm leaf rib and poke it into nest openings to collect the insects. The peak consumption of ants and termites occurred during the early part of the rainy season, when the winged reproductives (alates) of both kinds of insects swarmed. At this time, both men and children joined the women in collecting these foods.

While women were the primary harvesters of edible insects, men actively participated in collecting palm weevil larvae of *Rhynchophorus* spp. (figure 3.2). Unlike the other insect foods, which were foraged, this resource was semi-cultivated. The beetles predictably laid their eggs in fallen trees, so returning to a log two to three months after it was cut down would almost guarantee that weevil larvae would be present when the logs were split open. Men collected these insects more than any other because they visited these caches of larvae while they were away from the village on hunting or fishing trips.

The Yukpa-Yuko of Colombia and Venezuela

The Yukpa-Yuko are indigenous people from the border of Colombia and Venezuela who primarily practiced horticulture, supplemented by hunting, fishing and gathering. Ruddle (1973) detailed the edible insect practices of the Yukpa-Yuko, noting that insect foraging became increasingly important because of the destruction of the forest and its subsequent replacement with savanna and low regrowth vegetation. When large game became depleted and it took more time and energy to locate the animals, insects became the most reliable source of animal foods for the Yukpa-Yuko. Additionally, harvesting insects that were pests to the crops of these horticulturalists served a dual purpose as both food provision and pest control.

Women and children gathered short-horned grasshoppers (*Conocephalus* spp.) from agricultural fields using only their cupped hands. Occasionally during the peak grasshopper season, men set fire to the dry grass in the valley, pushing the insects to ridge tops, where women and children waited to beat down the swarms with large branches. Ants of the genus *Atta* were agricultural pests. On rainy mornings, a group from the village would make their way to a designated ant nest where the entrance would be dug out to create a moat for collecting the heavy rains. As the nest flooded, the moat prevented the ant colony from escaping. The Yukpa-Yuko selectively captured and killed the large females that were bloated with eggs in their abdomens, preventing the propagation of this agricultural pest. At the end of the day, the Yukpa-Yuko would return with several small baskets full of these ants to be eaten—sometimes raw but often roasted.

Beetles were consumed in small quantities but were highly valued, especially the "meaty" rhinoceros beetle (*Podischnus agenor*), which women and children sought at night during the early rainy season. Other beetles such as those of the genus *Caryobruchus* were also consumed, but usually in their larval stage. To collect these, women gathered palm nuts that had fallen to the forest floor and begun to rot; it was at that time that large beetle larvae could be extracted from the nut to be prepared in a variety of ways. Non-beetle larvae were also collected from nearby streams when available: women and children collected the larvae of the dobsonfly (*Corydalus* spp.) by searching under river rocks. They also found caddisfly larvae (*Leptonema* spp.) under rocks in strong currents and collected soldier fly larvae (*Chrysocholorina* spp.) from pools of water that had become disconnected from the flowing stream.

The Ngandu, Central Africa

The Mongo peoples are a horticultural ethnic group that was traditionally dispersed across equatorial central Africa. The Ngandu subgroup resides in the Democratic Republic of Congo, formerly known as Zaire. Research conducted in 1975 by Takeda (1990) reported that this population primarily cultivated cassava, but other crops included sweet potatoes, bananas, and peanuts. They also relied on at least twenty-four species of wild plants for food. Men were primarily responsible for hunting and trapping game and clearing the forest to make way for crops. Women tended the agricultural fields daily and gathered the bulk of the insects.

Takeda identified over twenty species of insects that the Ngandu consumed. Of these, caterpillars, termites, and beetle larvae appeared to be of greatest value. Seasonally available caterpillars of the family Noctuidae were abundant from August to November. These insects took over their host trees in numbers so large that the sound of them eating could be heard from below. After they consumed the leaves, they fell to the ground, where women and children would collect them. Younger women and children excelled at the task; they could collect up to 3,000 caterpillars in just a few hours.

The Ngandu consumed termite soldiers of the genus *Macrotermes* year-round, but they ignored the flying alates when they were available seasonally. They sometimes found beetle larvae of *Rynchophorus phoenicis* and *Augosoma centaurus* in trees that had been previously cut, but more often, they would cut down trees specifically to harvest the beetle larvae when they could hear their activity inside.

* * *

The studies discussed in this chapter represent some of the only available data on edible insect use in foraging populations that rely on traditional methods for their subsistence. Although the data is limited, some generalizations can be made. Each group consumed multiple insect species and these varied across populations. However, social insects such as ants and termites and seasonally abundant insect larvae from moths and beetles seem to hold the highest value across foraging cultures. Most notable is that women, more than men, do the foraging for insects. Children, who often accompany their mothers, also forage for insects more than the adult men in the community. It can be difficult to assess how much these insects contribute to the diet, especially when they are consumed as snacks while foraging. Researchers may be able to quantify foods that are prepared in meals, but it is more difficult to determine how many of these insects are eaten as they are found. This elusive portion of the diet largely belongs to women. In these studies, women ate more insects than men, and there is a strong possibility that there is an evolutionary explanation for this pattern.

4

Nutrition and Reproductive Ecology

I often find myself being asked in casual conversation what I study. My answer has long been "I study the evolution of the human diet and focus particularly on eating insects." To my surprise, *eating insects* has been often misheard as *eating and sex*, to which people often respond with "ooh-la-la" or a statement that they also fancy themselves to be experts on these topics. I reply, "No, eating *bugs*, but eating and sex is the evolution part." Over the course of human evolution, natural selection has been the strongest force in shaping our lineage. Nutrition and reproduction are inextricably linked in natural selection: without food or sex, our ancestors could not have survived and we would not be here today. Through the lens of natural selection, we can begin to look at eating insects as a behavior that ensured that women had enough food to meet the nutritional demands of childbearing and child-rearing.[1]

Evolution by natural selection is the differential reproduction of individuals that are best able to thrive in their environments. Individuals bearing beneficial traits tend to have more offspring thereby passing down higher proportions of these traits to subsequent generations. Natural selection acts most strongly on processes that increase chances of survival early in life, from conception through weaning, since producing healthy babies is critical to reproductive success (Stanford, Allen, and Antón 2011). In order to be successful in passing down one's genetic material to the next generation, many different elements need to be in place: hormones and other physiological processes associated with fertilization and conception need to be regulated, structures in the offspring need to be properly developed, and the mother must have the energy to gestate and ultimately raise the offspring. Simply put, there is a need for regulation, structure, and energy. These are the three key roles that nutrients fulfill in our bodies. This is why nutrition and evolution are tightly connected.

Roles of Nutrients in the Human Body

At the most fundamental level, nutrition is related to natural selection because people need enough energy to survive and reproduce. The energy-yielding nutrients include carbohydrates, fats, and protein. These nutrients are also commonly referred to as macronutrients because our bodies require them in large amounts. Protein is used as an energy source only when energy cannot be obtained from carbohydrates or fats or when protein intake exceeds the body's requirements. Being able to pull energy from protein is a mechanism that can be used until the preferred energy-yielding nutrients become available again.

Protein and fat are also important in structural roles. Lipids are important at the cellular level because they make up the membranes that surround cells, but protein is more associated with the structure of the body; it makes up most of our soft tissue. Even bone, which we think of as being composed of minerals such as calcium, has an embedded protein framework. Since reproduction is essentially the building of babies, these structural elements need to be present in order to build strong offspring that have the potential to reach adulthood.

Proteins are comprised of smaller units known as amino acids, which can be thought of as the building blocks of life. Our DNA, which is the coded information that gets passed down from parent to offspring, arranges amino acids and forms proteins, including those that direct the development of offspring. The genetic code includes twenty different amino acids, of which nine are known to be essential; our bodies cannot produce them from simple molecules and they must be derived from dietary sources. Animal-based foods are complete sources of these essential amino acids. Vegetarians and vegans are often concerned about meeting protein requirements, but their vigilance would more properly be directed toward receiving adequate amounts of each essential amino acid. This requires combining different plant-based resources. When these dietary requirements are not met, DNA cannot function properly (Stanford, Allen, and Antón 2011).

When we consume sources rich in these amino acids, our bodies break down the food into its constituent parts, making these amino acids metabolically available. All of the chemical processes that occur in the body are collectively known as metabolism, and these processes, which involve all of the nutrients, must be regulated to maintain the body's homeostasis. The different nutrient categories each help maintain homeostasis. For example, lipids produce the hormone estrogen, which regulates the reproductive cycle; the protein leptin regulates the size of fat stores; carbohydrates attached to circulating proteins signal when the liver should remove those proteins; B vitamins regulate the use of macronutrients for energy; minerals known as electrolytes, such as sodium, regulate blood volume; and water regulates body temperature through sweating.

Micronutrients are the vitamins and minerals the body requires in order to function properly. However, these need to be consumed only in small amounts. The B vitamin folate, or folic acid, is essential for regulating DNA synthesis and cell division. Any process that involves rapid cell proliferation, such as the development of embryos and fetuses, requires folic acid. For instance, early in pregnancy, the neural tube, which eventually becomes the central nervous system, forms through a complicated process of perfectly timed cell development. In the absence of folate, this process may be interrupted, causing the neural tube to form improperly, which can result in life-threatening birth defects such as spina bifida. This is why folate is a key ingredient in the prenatal vitamins obstetricians prescribe. Over the course of human evolution, we evolved mechanisms to help meet folic acid requirements, including having darker skin in equatorial areas to protect folic acid in the body from degrading when exposed to high UVA radiation (Jablonski 2013).

Another aspect of skin color evolution—the selection for lighter skin at higher latitudes—is tightly linked to nutritional needs. The intensity of UV radiation decreases with increasing distance from the equator. At latitudes distant from the equator, UVA rays are not strong enough to degrade folic acid, but the intensity of UVB rays, which synthesize vitamin D, is also much lower. A balance of skin color and strength of sunlight is necessary to produce the appropriate levels of vitamin D (Jablonski 2013). A recent cross-cultural study of vitamin D levels of people native to their region demonstrated that everyone maintains about the same levels of vitamin D in their body (Hagenau et al. 2009). These values, however, suggest that everyone has a lower level of vitamin D than what is recommended. Clothing and indoor life are likely reducing the exposure of skin to UVB rays in people around the world. Vitamin D can be obtained from only a few food sources: liver, egg yolks, and fatty fish such as tuna. Increased dietary intake of vitamin D through these natural sources or through artificially enriched food sources is important for avoiding deficiency in our modern lifestyles, which is critical since vitamin D is responsible for the intestinal absorption of minerals such as calcium and iron.

Calcium is the most abundant mineral in our bodies; its main reserve is in bones and teeth, where it has a structural role, but it is also important in the blood and soft tissues, where it regulates several metabolic processes. Calcium levels in the body are well regulated; bone and blood calcium levels are maintained within narrow limits. Active absorption of dietary calcium is regulated by vitamin D and its intestinal receptors based on the body's needs. When levels of absorbed calcium are low, stores of calcium from bone will be drawn into the blood. Because of the adaptation of these mechanisms, severe nutritional deficiency of calcium is rare, but chronic low intake will lead to reduced bone mass, placing the individual at risk for osteoporosis and bone fractures.[2] Milk and milk products are excellent sources of calcium, but if these are not consumed, intake of calcium may be low. Other sources of calcium are thus critically important to populations that do not consume milk. For our ancestors, this need was met by getting small amounts from a variety of foods. Plant foods, such as leafy green vegetables and nuts, contain calcium, but often these foods have anti-nutrients such as oxalic acid and phytic acid that inhibit calcium absorption. These compounds can also interfere with iron absorption.

Iron is a trace mineral micronutrient that functions as a component of a number of proteins. Most notable among these proteins is hemoglobin, which is important for transporting oxygen to tissues throughout the body. In adults, anemia resulting from iron deficiency can cause impaired cognitive performance, but in developing children, deficiency can lead to impaired cognitive development. Iron-deficiency anemia during pregnancy is associated with preterm delivery, low birth weight, and infant and maternal mortality. Pregnancy depletes maternal iron stores, but after birth, a woman's needs are greatly lowered and she is often able to recover before her next pregnancy, despite any losses that may occur via menstrual blood.[3] Iron is more bioavailable from meat than from plant-based foods, so the omnivorous condition is beneficial for ensuring that these stores get replenished.

Another nutrient that is most easily obtained through omnivorous diets is vitamin B_{12}. This vitamin is synthesized by bacteria, so it does not exist in any plant product that has not been fermented. The richest source of vitamin B_{12} is organ meat, but it is also found in significant quantities in muscle meat, shellfish, chicken, and fish and in lesser amounts in eggs and dairy products. Only a small amount of B_{12} is necessary in the diet, but it plays a key role in DNA synthesis, neurological function, and the formation of red blood cells.

Water is also a nutrient. Even though it is inorganic and does not yield energy, it is necessary for survival and must be obtained from outside sources. Water makes up more than 60 percent of the human body. It is estimated that a human can survive for a month or more without eating but only a week or so without drinking water (Janiszewski 2015). We rely on water to make everything else work; it makes up 90 percent of our blood, which we need to transport nutrients, oxygen, hormones, and other resources throughout the body. Without the circulatory system, nutrients would not be able to fulfill their roles. Water is also important in regulating our body's temperature and pH level to maintain homeostasis. Humans are able to sweat from head to toe, which suggests that there was an evolutionary advantage to being able to cool off efficiently. Sheens of sweat that cover our body lead to evaporation and transfer heat from our skin to the air. The refinement of this cooling mechanism likely evolved over a million years ago, when the ancestors of all humankind lived on the continent of Africa. Being able to travel by day to collect food or to hunt without suffering heat stroke must have been a great advantage. The important catch in the system, however, is that water must be replaced, ideally at the rate at which it is lost, to avoid the negative effects of dehydration. Although drinking water is an easy way to maintain hydration, people get much of the water they need from food.

The Nutritional Requirements of Women

As the bearers of children, women invest much more in their offspring than men do, from gestation and lactation to childcare; a man may be reproductively successful without ever meeting his offspring. For these reasons, women's reproduction is much more limited than that of men because of the investment that comes with each child, while men can have as many children as they have reproductively viable sexual partners. A simple search of Guinness World Records demonstrates the disproportion of reproductive potential between the sexes. The greatest officially recorded number of children born to one mother is sixty-nine. The wife of Feodor Vassilyev (1707–ca. 1782), a peasant from Shuya, Russia, completed twenty-seven pregnancies in which she gave birth to sixteen pairs of twins, seven sets of triplets and four sets of quadruplets. This account was corroborated by many sources. Although sixty-nine children may seem astonishing, it pales in comparison to the estimates given for men. Although there is currently no complementary entry for most prolific father in Guinness World Records, previously Guinness noted that Ismail Ibn Sharif, the reigning sultan of the Alaouite Dynasty in Morocco from 1672 to 1727, had fathered 867 children (Stanford, Allen, and Antón 2011). Of course, this number is impossible to verify, but it highlights the increased reproductive potential of men compared to women. Estimates for the number of children fathered by Genghis Khan range from over 1,000 to possibly upward of 2,000 children. Genetic evidence suggests that 16 million men today can trace their ancestry back to Genghis Khan (Balaresque et al. 2015).

Differential investment in offspring means that the sexes have different needs for reproductive success. Men need adequate nutrition in order to have the energy to find a potential reproductive partner, the strength to woo that partner through whatever means are expected in their society, and to regulate physiology for successful sperm production and ejaculation. Women have analogous needs for the process of finding a mate. They also have requirements that affect the reproductive success of both parents, since the nutrients from a woman's diet direct the development of the child. Women who are pregnant or lactating not only have increased energetic requirements but also an increased need for nutrients that have structural and regulatory roles during pregnancy. The amino acids necessary for building the structures in the baby and the vitamins and minerals that regulate the baby's developmental processes all need to be available.

Many of the dietary requirements of reproductive women increase beyond those of men (table 4.1). Men are known to have larger protein requirements than women because they carry, on average, larger amounts of muscle mass. For men, the recommended dietary allowance (RDA) of protein is 56 grams per day. For women, the protein requirement is 46 grams per day, but this increases to 71 grams when they are pregnant or

Table 4.1. Recommended daily allowance (RDA) of selected nutrients for men, nonreproductive women, pregnant women, and lactating women

Nutrient	Men	Nonreproductive women	Pregnant women	Lactating women
Protein (g/day)	56	46	71	71
Carbohydrates (g/day)	130	130	175	210
Iron (mg/d)	8	18	27	10
Folic Acid (mg/day)	400	400	600	500
Calcium (mg/day)	1,000	1,000	1,300	1,300
Vitamin D (IU/day)	600	600	600	600
Vitamin B_{12} (mcg/day)	2.4	2.4	2.6	2.6

Sources: RDAs for macronutrients: Institute of Medicine of the National Academies (2005); RDAs for trace minerals: Food and Nutrition Board, Institute of Medicine (2011).

lactating. Women who are pregnant or lactating also need more energy than men. For both men and women, the RDA for carbohydrates is 130 grams per day, but this increases to 175 grams when women are pregnant and 210 grams when they are lactating.

It is also important for reproductive women to increase their intake of critical micronutrients. Women's RDA for iron is 18 milligrams a day. That value increases to 27 milligrams when a woman is pregnant but reduces to about 10 milligrams when a woman is lactating. Once the woman stops lactating, her daily requirements increase back to 18 milligrams to support her returning menstrual cycle. For men, the RDA for iron is 8 milligrams of iron a day, with a slight increase to 10 milligrams when young men are going through puberty. Calcium requirements increase during periods of rapid growth, which include infancy, pregnancy, and lactation. Pregnant and lactating women need 1,300 milligrams per day of calcium compared to 1,000 milligrams for men and nonreproductive women. Requirements for vitamin D, however, do not increase, because this nutrient is always regulating calcium absorption and the increased calcium women need during pregnancy does not strain this system. Lastly, pregnancy can lead to the decrease of maternal B_{12} due to the rapid transfer of the nutrient to the fetus. The required amount of B_{12} for both men and women is only 2.4 micrograms a day while the RDA increases to 2.6 micrograms per day for pregnant women and remains that high during lactation to ensure that the infant's needs are met.

Although it is in both parents' interest to ensure that the mother's nutritional requirements are met so that the baby is given its best chance of survival, the amount of investment each is willing to make in this effort may be different.

Theories of the Sexual Division of Labor among Hominins

In foraging cultures, men and women acquire foods differently (Brown 1970; Bird 1999; Marlowe 2007). Men tend to take higher risks, going after higher-reward resources such as large game, while women spend their time obtaining more reliably available resources such as tubers, nuts, and insects that can be collected while also tending to young children. The widespread nature of this gendered foraging pattern leads to the question of the potential evolutionary origins of the sexual division of labor.[4]

One hypothesis for the evolution of differential resource procurement by sex is the conflict model. This states that male and female physiologies have different potential conflicts with their environments, meaning that men and women have different nutritional needs and thus use different strategies to procure resources to optimize their fitness. An alternative hypothesis that originates from studies of monogamous birds is the cooperative provisioning model (Lack 1968). This model posits a strategy whereby men and women target different resources and then bring them back home to feed their offspring. When men and women are not competing for the same resources they are able to more fully exploit the environment and provide nutritional variety for their families. Since having both sexes contribute to the well-being of offspring increases the reproductive success of both parents, this behavior is well suited for favor by natural selection.

A key component of the cooperative provisioning model is having a meeting place, such as a nest or a breeding ground for birds or a home base or camp for hunter-gatherers. Nonhuman primates do not utilize such spaces; they travel throughout the day in search of food and settle in new sleep sites each night. Young offspring accompany adults on these treks by using their hands and feet to cling to the hair of their mothers or other adult (figure 4.1). Although nonhuman primate fathers may carry infants and share food with them, which occurs most commonly in socially monogamous species where paternity is more certain (Lukas and Clutton-Brock 2013; Strier 2015), the cooperative provisioning model

Figure 4.1. Unlike other mammals, primates do not leave their young in nests and instead invest heavily in their care by carrying them throughout their infancy. Here, a chimpanzee mother named Tanga carries her infant at Gombe Stream National Park in Tanzania, November 2016. The grasping hands of young primates enable them to secure themselves to the hair of their adult caretakers. Photo by J. Lesnik.

envisions a different kind of social organization that is unique to the hominin lineage.

The question then is when in the last five million years of human evolution did social organization begin to resemble that of modern hunter-gatherers? When did fathers begin to bring high-risk resources back to base camps?

Lovejoy (1981, 2009) offers a controversial yet highly cited view; he proposes that the origin of the nuclear family, in which paternal investment was important to the success of offspring, can be traced back to 4.2 million years ago. His argument is based on canine tooth size, which is similar between the sexes in early hominins, but is often enlarged in male primates that exhibit high amounts of competition for access to female mates. However, this view is widely criticized, especially by feminist scholars (Walrath 2006). Although less dimorphic canine size has been correlated with reduced male-male competition in nonhuman primates (Leutenegger and Kelly 1977), in a number of primate societies, this pattern does not hold (Hrdy 1981). But the primary issue is that it is a huge leap in logic to extrapolate that hominin males were bringing resources back to feed nuclear families. Comparative morphology of dentition between hominins and nonhuman primates is a valid line of reasoning, but since no nonhuman primate society exhibits the level of provisioning Lovejoy suggests, his hypothesis does not hold up to scrutiny.

Wrangham and colleagues (1999) and Carmody and Wrangham (2009) offer an alternative timeline by crediting the evolution of the significantly larger brain size that we see in *Homo erectus* as the evidence that foraging strategies must have shifted to include fire and communal behaviors including bringing food back to be cooked and shared. The large adult brain sizes of *Homo erectus* suggest that babies underwent a longer time of postnatal brain growth, which left them helpless for longer periods of time (DeSilva and Lesnik 2008). Communal provisioning would have been important to the survival of *Homo erectus* babies, which suggests that male provisioning had evolved in the hominin lineage by about 1.8 million years ago.

Although incontrovertible evidence of controlled use of fire does not exist in the archaeological record for early *Homo erectus*, there was an increase in what are known as base camps. Accumulated stone tools and butchered remains of fauna define these camps and suggest that hominins transported scavenged and/or hunted meat to these areas (Bunn 1981). Transport, however, does not necessarily mean sexual division of labor. It is likely that these camps were first used as a way for hominins to defend valuable meat resources from carnivores (Rose and Marshall 1996). Cooperative provisioning may have evolved shortly after the initial use of such areas as a way of alternating who was defending resources and who was out looking for more.

In addition to the difficulties of identifying when cooperative provisioning behaviors arose during the course of human evolution, it is difficult to identify the exact benefit of the activity, especially in terms of male reproductive success. According to optimal foraging theory, animals will optimize how they acquire resources to gain the most benefit for the lowest cost. This theory, which Charnov (1976) originally proposed, is a model for predicting how an animal will behave when searching for food. It has been applied to modern human hunter-gatherers and modified accordingly over the years to account for cultural benefits that may come from expending energy. For instance, it is difficult to argue that large-game hunting, a risky activity with uncertain return (Bird, Codding, and Bird 2009), is an optimal foraging strategy. However, if participating in such behavior and being successful brings the individual higher status, then the benefit may be best calculated in terms of reputation and potential future matings. Therefore, the basic tenets of the cooperative provisioning, that it evolved because it is an effective way to provide nourishment to offspring, does not tell the full story of the benefits of hunting.

The conflict model and the cooperative provisioning hypothesis are not mutually exclusive. For hunter-gatherer men, hunting success is an important route to prestige in the community, and higher status correlates with increased reproductive fitness (Gurven and von Rueden 2010). Since successful hunting helps men secure mates, this pattern of division of labor arguably fits the conflict model. However, large game provides more food than any one individual can consume, and males who hunt large game receive benefits from provisioning their offspring, which simultaneously supports the cooperative provisioning model. For the mothers, cooperative provisioning reduces the energy they need to exert to feed their children and thereby allows them to have subsequent offspring sooner, increasing total fertility. However, it is possible that the resources that women forage best support their unique nutritional needs, which would support the conflict model as well.

If division of labor was entirely about provisioning, it would make more sense for a provisioning man to do what the women do and target reliable resources instead of ones that are highly variable and widely shared (Bird 1999). Although men may strive for status through hunting, in most foraging societies there is some overlap in the resources men and women acquire, and men may be better providers when the need arises. The Hadza foragers of Tanzania offer an example. Hadza men focus on hunting and on collecting wild honey, and the bee larvae within, while women dig wild tubers (the mainstay of the Hadza diet) and gather berries and baobab fruit. The tubers are rich in carbohydrates but offer little else (Schoeninger et al. 2001). Hunted and scavenged meat is the highest-quality food in the Hadza diet, but hunters are not successful every day and their acquisition rates vary seasonally. If no meat is brought home, nursing mothers who have increased protein requirements plus decreased mobility for procuring resources risk not meeting their nutritional needs. It is during this time that men will bring in resources more commonly foraged by women, but fathers are more likely to provide for their families than stepfathers (Hawkes, O'Connell, and Jones 2001; Vincent 1985; Marlowe 2003).

Abundant research on hunting and meat sharing in foraging societies demonstrates how men's foraging strategies exhibit tradeoffs between provisioning and prestige. The patterns are less clear for the insect portions of forager diets because the data are far less available. Most often, if insects as food are assessed at all, they are grouped together with other small gathered nonvegetarian resources, such as invertebrates, small lizards, eggs, and so forth. This broad category is represented in the Standard Cross-Cultural Sample (SCCS), a database that contains descriptions of 186 preindustrial societies along with corresponding coded behavioral data. Using this database, Marlowe (2007) identified only eighteen foraging groups that had data on collected small animal foods. Of these eighteen populations, there seemed to be an almost even split in the gendered division of labor: in seven of the eighteen groups, such foraging was done by men, and in eight of the eighteen groups it was done primarily by women (in three groups, foraging was done by both genders equally). The data in the SCCS for the !Kung San suggests that it is the men who almost exclusively gather small nonvegetarian items. However, Nonaka (1996), in a study that focused on insect consumption, found that women foraged for insects more than men and would even stop to eat termites for extended periods of time if they came across a productive nest. Men also collected insects, but women consumed more insects than men. In another example, Marlowe (2007) reported that among the Mbuti hunter-gatherers of the Democratic Republic of Congo, men and women gathered small nonvegetarian items in equal amounts. However, in a rather comprehensive list of animal foods the Mbuti consumed, only four items were listed as "only consumed by women and children": frogs, crabs, freshwater snails, and land snails (Ichikawa 1987). This evidence suggests that in the Mbuti group, there is indeed a difference in how the small animal foods are consumed. Additionally, the Mbuti exhibit numerous prohibitions on the foods they consume during different life stages, usually meat. These restrictions are meant to prevent diseases or disorders in susceptible individuals such as unborn children, newborns, and infants and there are many foods avoided by pregnant women and their husbands in order to protect their unborn babies. Women are prohibited from eating additional types of meat, such as the organs of duikers caught in nets, because it is believed that if women consumed these foods it would spoil the productivity of net hunting. Although it is not specifically documented, it is probable that during times of meat restriction, women take particular advantage of the small animal foods that are available to them. These foods would not only include the frogs, shellfish, and snails that only they are allowed to eat but also the locusts, termites, caterpillars, and larvae that have no restrictions. Unfortunately, the San and Mbuti are the only two of those eighteen populations for whom the insect portion of the diet has been detailed enough in the literature to tease it apart from the consumption of other resources.

The data on modern foraging populations that include food consumption by gender shows that women often consume more insects than men (Bodenheimer 1951; Nonaka 1996; Hurtado et al. 1985). Although this fits our understanding that women tend to focus on foods that are smaller in size and have lower risk of pursuit failure (Bird 1999), it is important to know what nutrition the insects provide if we are to test which model for the evolution of these gendered patterns is the best fit for this food source. If insects offer the nutrients most important to reproductive women—energy, protein, iron, calcium, folic acid, and vitamin B_{12}—then the conflict model may be more appropriate than the cooperative provisioning model for explaining why women disproportionately target this resource.

Macronutrient Contributions of Insects

As a food group, insects are generally regarded as nutritious; they are thought to be high in protein and other nutrients. However, the nutrition insects offer is highly variable, even between species of the same genus and among the castes within social species. The insects that humans prey on tend to fall into five of the insect orders: Isoptera (termites), Hymenoptera (ants, bees, and wasps), Orthoptera (locusts and crickets), Lepidoptera (butterflies and moths, usually in larval form), and Coleoptera (beetles). Below are summaries of the nutritional contributions of these insects. The values below are given as a percentage of the dry matter (DM) present in the insect when possible. Most insects contain 50–80 percent moisture, so the dry matter composes at most 50 percent of the insect. This is an important scale to keep in mind. Eating a single insect will not provide significant nutritional benefits; they must be consumed in larger quantities. Similarly, if these larger quantities of insects are consumed raw, they provide water as well, which may be useful in dry climates.

Macronutrients

Termites (Isoptera)

Termite reproductive castes—the alates and nymphs—are available only at certain times of the year and are a desired food resource for many populations in Africa, Australia, and South America (Bodenheimer 1951). These termites tend to have the highest fat content of all the termite castes. However, termite soldiers, which are available year-round, offer the highest protein values within a termite species. The termite soldiers of *Syntermes* that the Tukanoan in the Amazon basin consume contain about 59 percent protein and 5 percent fat. These values were taken from dry-roasted termites, so the values are not reported as percent DM, but they should be relatively close (Dufour 1987). Although there is no comparable value for the alates that the Tukanoan consume seasonally, the fat content likely is significantly higher. This is suggested by an analysis of *Syntermes dirus* alates that yielded 22 percent DM (Redford and Dorea 1984). Today, most people consume termites of *Macrotermes*; one example is *Macrotermes subhyalinus*. The nutritional offerings of these termites are slightly different than those of *Syntermes*. The soldiers of *M. subhyalinus* are extremely high in protein (74 percent DM) but low in fat (3 percent DM). The alates, however, have much more fat (51 percent DM) but still offer significant protein (41 percent DM) (O'Malley and Power 2012).

Ants (Hymenoptera)

Similar to the nutritional values across the caste system in termites, ant nymphs and adult reproductives tend to be higher in fat than the rest of the castes, while the soldiers are the highest in protein. In weaver ants (*Oecophylla* sp.), like those the Nyishi and Galo peoples of northeast India consume (Chakravorty, Ghosh, and Meyer-Roschow 2011), the soldiers and workers that are available year-round contain 12.5 percent DM fat and 61.3 percent DM protein, while the seasonally available reproductives contain 39.8 percent DM fat and 50 percent DM protein (O'Malley and Power 2012).

Locusts, Grasshoppers, and Crickets (Orthoptera)

Insects of Orthoptera are not social, so individuals of the same species offer roughly the same nutrients because they are not differentiated in a caste system. Even across the order, protein yields in these species are relatively consistent. Most locusts, grasshoppers, and crickets contain about 50–70 percent DM protein, similar to the amounts in termite and ant soldiers. However, fat content is more variable and recorded values cover a broad range. Some locusts have been reported as containing as much as 50 percent DM fat (*Locustana* sp.), while other genera are relatively low in fat at only 4.5 percent DM (*Oxya* sp.). Thus, these insects appear to be a reliable protein source, but the fat contribution depends on which species is consumed (Leung, Busson, and Jardin 1968).

Beetle Larvae (Coleoptera)

Beetles are commonly consumed in their larval form. The large grub of the palm weevil is a highly sought-after food in many tropical regions. In general, the larval stage contains more fat than the adult form, but for palm weevils of the genus *Rhyncophorus*, the late larval stage has both the most fat and the most protein (Omotoso and Adedire 2007). The smoke-dried palm larvae that the Tukanoan in the Amazon basin consume contain 55 percent fat and 24 percent protein (Dufour 1987). Multiple studies have analyzed the nutritional content of the African palm weevil (*R. phoenicis*). The values for fat and protein have a broad range, possibly due to differences in techniques of analysis, but the average fat content across the studies is sizable at 50.2 percent DM. Protein is also well represented at 30.4 percent DM (Omotoso and Adedire 2007; Onyeike, Ayalogu, and Okaraonye 2005; Banjo, Lawal, and Songonuga 2006).

Caterpillars (Lepidoptera)

Caterpillars are the larval-stage insects of moths and butterflies. Although larval insects tend to offer significant amounts of fat, values for caterpillars show that they are quite variable across this order. In an analysis of eleven species, average fat content was 27 percent DM but the range was large, from 7 to 77 percent. However, the protein content of caterpillars is more consistent; these same eleven species averaged 46 percent DM protein with a range of 15–60 percent DM (Ramos-Elorduy et al. 1997). The witchetty grubs of Australia have a large range of recorded nutritional values because they represent multiple species of moth larvae from the Cossidae family (Yen et al. 2017). It is best to have values for the exact species of caterpillar consumed when estimating its nutritional contributions. For instance, the hawk moth caterpillar, *Herse convolvuli* (family Sphingidae), which the San in southern Africa regularly consume (Nonaka 1996), has over 50 percent DM protein and 21 percent DM fat (Shao-jun and Yong-huang 1997; Li et al. 2006).

Micronutrients

The micronutrient content of insects is very variable across their orders. Unlike fat and protein, which must be present in some amount since they constitute the structural components of insect bodies, the presence or absence of micronutrients in an insect is likely related to its diet. For instance, termites that feed on soil have an abundance of micronutrients compared to the species in the order that forage for wood or grass. Unfortunately, much less data is available for insect micronutrients than for insect macronutrients.

In the available micronutrient datasets, insects of the orders Orthoptera and Coleoptera appear to be quite rich in folic acid (Rumpold and Schlüter 2013). Insect levels of iron are highly variable, but it seems that certain species within Hymenoptera and Isoptera are the best sources (Bukkens 1997). Edible insects seem to always contain at least some iron. For instance, palm weevil larvae (*Rhyncophorus*) contain 14–30 milligrams of iron per 100 grams of DM (Rumpold and Schlüter 2013). A pregnant woman needs 27 milligrams of iron per day, and although she would have to eat at least 200 grams of palm weevil larvae to meet her daily requirement, eating even some these larvae could still be an important dietary source of iron. All insects have some calcium as well, but generally in low amounts. The average calcium content in recorded edible insects is about 50–175 milligrams per 100 grams DM, not a significant contribution toward the human requirement of 1,300 milligrams daily (Bukkens 1997; Rumpold and Schlüter 2013). But for foragers, this insect consumption might be important if they needed to obtain calcium from a wide variety of nondairy foods. As for vitamin B_{12}, not all insects contain the nutrient, but termites contain high amounts. Termites house large numbers of bacteria in their guts that are important for digesting their cellulose-dense diets, and these bacteria likely also synthesize vitamin B_{12} (Wakayama et al. 1984).

Humans have RDAs for the amino acids that make up protein. Insects tend to provide some amount of each of the essential amino acids. Of the essential amino acids, tryptophan tends to have the lowest value across the board for edible insects. However, tryptophan is abundant in a number of plant and animal foods. Seeds, nuts, and red meat tend to be good sources, so if some of these foods are present in the diet in addition to insects, tryptophan deficiency would not be a problem (Rumpold and Schlüter 2013).

Essential fatty acids are similar to amino acids in that they are required for metabolic processes in the human body and must be acquired through food since we are unable to synthesize them ourselves. Only two fatty acids are known to be essential for humans: alpha-linolenic acid (an omega-3 fatty acid) and linoleic acid (an omega-6 fatty acid). The fatty acid contributions of insects are variable but generally align with the offerings of fish and poultry. Rather high amounts of alpha-linolenic and linoleic acids have been reported in some insects. Insects that belong to Lepidoptera consistently offer high amounts of alpha-linolenic acid, and levels of linoleic acid appear to be high in insects that belong to Orthoptera (Rumpold and Schlüter 2013).

Limitations of the Current State of Our Knowledge of the Nutritional Values of Edible Insects

These nutritional profiles indicate that insects offer many nutrients and can be an important source of the specific nutrients that reproductive women need most. However, the available data on edible insect nutrients come from nutritional analyses of the insects themselves, and this does not indicate anything about the bioavailability of these nutrients to the human body. Simulations of how insect foods interact with the human gut must be conducted in order to determine what our bacteria and enzymes are able to extract from the insects and how much passes through our system without being absorbed.

One of the biggest areas where we lack knowledge is how our digestive system handles chitin. Chitin is a polysaccharide that is reinforced with hardened proteins and minerals such as calcium carbonate to create the exoskeleton of insects. Some nutritional analyses may indicate values obtained from the exoskeleton, but we do not know how well our bodies are able to metabolize this part of the insect. Humans possess the enzyme chitinase and chitinase-like proteins that can break down chitin, so in theory, the nutrients bound in the chitin should be available to us, but we currently do not have good data about whether this is the case (Janiak, Chaney, and Tosi 2017). I expect this area will be a research priority as interest in edible insects continues to grow.

* * *

Other factors beyond nutrition may affect how foods are collected, such as the potential to gain notoriety and status. For women, the benefits of having higher status are not as reproductively advantageous as they are for men. Instead, in foraging populations, a woman's best chance for reproductive success is to secure her nutritional needs so that her babies can develop into healthy children. One way that women may be able to do this is to cooperate with men, targeting different resources so that when taken together, all nutritional requirements are met. Support for the cooperative provisioning model would come from data suggesting insects offer important nutrients that are not found in the resources men procure. However, the data in this chapter demonstrate that this is not the case. Insects are an animal food and offer similar nourishment as the hunted game men target. Thus, the conflict model may be the best fit.

The question of when this pattern evolved remains, though. If the data had suggested the cooperative provisioning model, then it could have been stated with some authority that the pattern evolved with the genus *Homo.* It is with the evolution of *Homo erectus* that large-brained babies would almost certainly have had low survival rates if there were not some sort of additional provisioning beyond what the mother could provide. However, since the data suggest women may be targeting insects in order to ameliorate the nutritional conflicts of their reproductive physiology, then this may be a sex-based difference that goes back further than the origins of the nuclear family, with the possibility that it was present in the earliest hominins. One way to test this is to see if it is reasonable to reconstruct this pattern for the last common ancestor of humans and chimpanzees, and to do that, we have to look to insect eating by our closest living relatives, the nonhuman primates.

5

Insect Eating in Nonhuman Primates

One of my current research goals is to understand why people and chimpanzees have different preferences for the termite taxa they consume. Humans around the world eat termites from more than a dozen genera. In contrast, chimpanzees, our closest living primate cousins, eat termites of one genus, *Macrotermes*, almost exclusively. Chimpanzees' strong preference for these termites has led me to search for underlying chemical structures in the taxa they do not prefer that may deter them from consuming them (but that humans may bypass via cooking or other processing). For this ongoing work, I have collected termites from three sites in Africa, one of which was Gombe Stream National Park in Tanzania, the chimpanzee site where Jane Goodall first observed chimpanzees using tools to extract termites from their nests. While at Gombe, I spent a few days out with the chimpanzees so that I could observe some termite foraging. Of the handful of foraging bouts I was able to witness, one stood out for me. When I arrived at the termite mound, two chimpanzees were present: an adult female called Glitter and her juvenile daughter, Gossamer. Glitter was searching the surrounding vegetation for the right materials to prepare a tool and Gossamer was standing by quietly. All of a sudden Gossamer started screaming, loud vocalizations that clearly indicated that she was upset. I was concerned initially that as a new visitor to the site, I was somehow disturbing her, but Glitter did not seem fazed by my presence or the antics of her child and continued to search for a tool. Once she found a tool and settled in at a spot on the termite mound, Gossamer crawled into her lap and began to nurse (figure 5.1). I had just witnessed a hungry child's temper tantrum.

Figure 5.1. Adult female chimpanzee Glitter nursing her not yet fully weaned 4.5-year-old daughter Gossamer while she fishes for termites (*Macrotermes subhyalinus*) at Gombe Stream National Park in Tanzania, November 2016. Photo by J. Lesnik.

Just like people, chimpanzees use tools to acquire termites and females can access this resource with their young in tow. After Gossamer was finished nursing, she grabbed her own tool, although using less consideration than her mother, and displaced Glitter from her fishing spot. Glitter relocated to another spot and continued at a different opening in the mound, only to be displaced again by Gossamer a short while later. Glitter and Gossamer's interactions during the half hour I spent with them demonstrated in two ways how they value this resource: 1) chimpanzee mothers will eat termites while simultaneously tending to the demands of their young; and 2) mothers create opportunities for their offspring to independently procure the resource for themselves. In this chapter, I review the importance of edible insects to the taxonomic order Primates and investigate whether nonhuman primates that eat insects as a supplement to their diet exhibit patterns of sex difference in this resource's procurement that are similar to those seen in modern human foraging groups.

Introduction to the Order Primates

Members of the order Primates are united by a suite of features that include stereoscopic vision, the presence of a clavicle, and grasping hands. Forward-facing eyes give primates binocular, stereoscopic vision that makes depth perception possible. Emphasis on the sense of sight in primates is also evident in their reduced sense of smell compared to other mammals. The costs of reduced olfaction and of limiting the field of vision by moving the eyes away from the sides of the head suggest the importance of depth perception to the earliest primates. Increased mobility of the forelimb was also important to these early primates. The clavicle positions the shoulder joint away from the thorax, giving the arm a greater range of motion. The hands of primates generally have five grasping digits with nails instead of claws and sensitive pads on the fingertips. Often, the pad on the thumb can reach and contact the pad of at least one other digit, which is the definition of an opposable thumb. Today's primates take great advantage of these traits when moving about in the trees, but they may not have evolved for that reason.

Although primates are expert tree climbers, an arboreal hypothesis for the origin of primate traits is unsatisfactory because other arboreal mammals, such as squirrels, are accomplished climbers and do not possess these traits. An alternative hypothesis, the visual predation hypothesis, suggests that these traits evolved to enable primates to catch insect prey (Cartmill 1974). Although this hypothesis can account for the suite of primate traits, it discounts the reliance on olfaction that the ancestral primate likely had. The last common ancestor of all primates most closely resembled today's strepsirrhines, one of two major branches in the Primate order (Bloch and Silcox 2006) (figure 5.2). Strepsirrhines include lemurs and other primates that have wet noses and projecting faces, traits that emphasize olfactory over visual cues when searching for food. The angiosperm hypothesis, sometimes called the mixed-diet hypothesis, emerged as a way to explain traits such as grasping hands while also accounting for a keen sense of smell (Sussman 1991; Nekaris 2005; Sussman, Rasmussen, and Raven 2013). This model proposes that the primate suite of features evolved so primates could spend more time in trees in response to an increase in the availability of angiosperms, or flowering plants. The flowers and fruits of these plants are good to eat and their aromatic sweetness attracts edible insects. Hands and feet that grip and eyes that provide excellent depth perception would have enabled primates to visit the terminal branches of trees to reach these foods that their noses helped them locate. For most flowering plants, both reproduction and dispersal require the cooperation of animals, often insects or birds. Around 70 million years ago, the number of these plants increased to about 80 percent of the vegetation. They co-evolved with an increased number of pollinators and seed dispersers, potentially including the small mammals that are ancestral to all primates.

Soon after the first true primates evolved, the two major branches, or suborders, emerged: Strepsirrhini and Haplorhini. Named for the differences in their noses, haplorhines include the dry-nosed monkeys, apes, and tarsiers. Small-bodied tarsiers resemble the earliest haplorhines, but the suborder has diversified greatly since its origin. New World monkeys are relatively small in body size compared to Old World monkeys and apes. They are arboreal quadrupeds that rarely come down from the trees, where they move about the tops of branches on all fours. Old World monkeys, which include baboons and macaques, have larger body sizes than their New World counterparts and spend considerable time on the ground, sometimes living in more open savanna environments. Apes, the largest primates, often spend a lot of time on the ground, although only a few populations of chimpanzees live in open savannas. In order to feed in the trees at this size, they have evolved suspensory behavior that enables them to hang below the branches so that they move with the branch that is carrying their weight instead of falling off it.

In addition to the suite of traits that enables primates to successfully find food in the trees, they all have a relatively large brain compared to other mammals of the same body size. With these large brains come

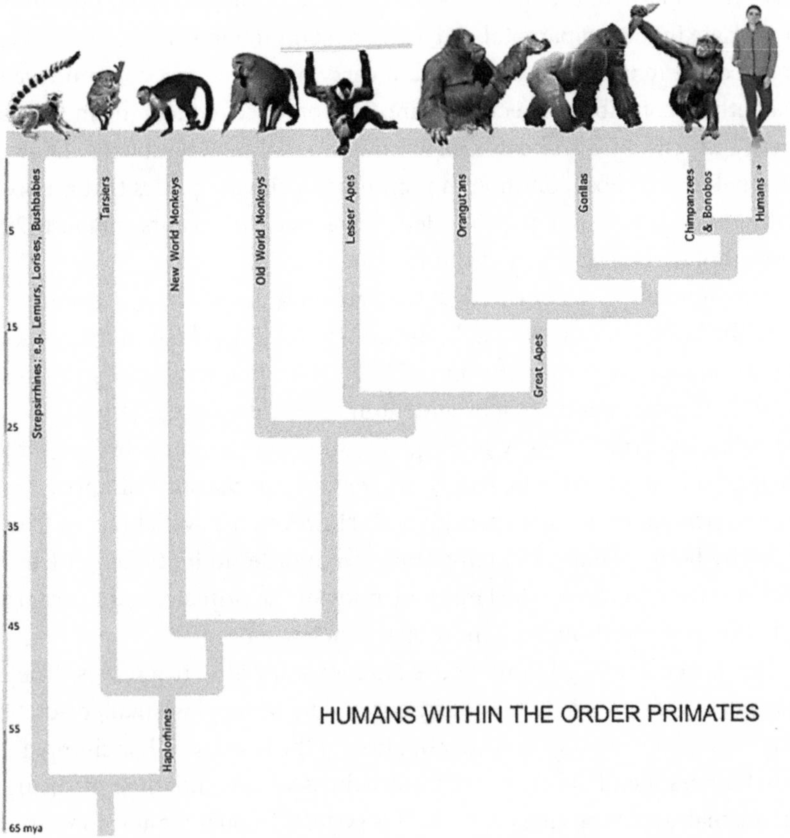

Figure 5.2. Phylogeny of the order Primates indicating the relationship of humans to the rest of the major branches.

correlated behaviors that are seen across the order, including sociality and an increased investment in offspring.

Most primate mothers have one infant at a time; only a few New World monkey species bear twins. All animals who bear single offspring invest more in their progeny than animals that birth multiple offspring. Instead of playing the odds by having large litters, these animals invest in the success of one offspring at a time. Because of their large brains, primate offspring develop more slowly than other mammals. This increases the amount of time mothers must contribute to their care. Primate infants are not left alone in nests, so they not only need to be fed, they also need to be carried everywhere the mother goes. Because of the increased demands of primate infants, biparental care is more common in primates than it is in any other mammalian order. Infant carrying is the most common way that fathers contribute directly to care. This practice reduces the mother's energy expenditure and can shorten the interval between births. Direct paternal care is most common in nonhuman primate species that are socially monogamous, or pair-bonded, a practice that occurs in about 29 percent of species (Lukas and Clutton-Brock 2013).

Primate mating systems reflect the ecological niche the species occupies. The body size of individuals, the number of individuals in the group, and the amount of fruit in the diet all affect the amount of space primates need. This pattern is especially evident in the great apes, the largest-bodied primates. Fruit is a very high-quality food source and primates with the largest brains tend to be frugivores, or fruit specialists. Fruit provides over 50 percent of the diet of frugivores. However, the availability of fruit is patchy both spatially and temporally, so in order to find enough fruit, primates need to travel. The larger the body of the primate, and the more individuals in the group, the more space they will need.

The lesser apes—gibbons and siamangs—are also frugivores. They manage by being socially monogamous and by occupying small, discrete ranges as family units so they do not outstrip the resources. For chimpanzees, females spend little time with individuals who are not their offspring but the males are very gregarious. In this system, females are able to spread out to secure the resources they need, while the socially bonded males work together to defend the large territory that encompasses the females and other resources. In contrast, orangutans, who are also large-bodied frugivores, are solitary. Unlike polygamous chimpanzees or gibbons and siamangs, which are essentially monogamous, orangutan males monopolize access to a few females by defending a territory around the females' core areas. Orangutans exhibit large degrees of sexual dimorphism. Males who defend territories are more than twice the size of females, a pattern that evolved because of the extreme competition among males to secure an area. Gorillas are similarly dimorphic, but unlike the other apes, they are folivores, not frugivores; they specialize in leaves rather than fruit. Because their diet is high in low-quality foods, gorillas maintain their large brain and body size by moving very little and thus not expending any more energy than is needed. This diet also enables females to group together and be social. A gorilla group will include one adult male that is easily identifiable by his large body size and characteristic silver back. This male defends his position against outsiders.

While leaves are an abundant, protein-rich, and widespread resource that takes extra time to digest, edible insects are a quickly metabolized source of protein that is spatially and temporally dispersed. Because of this dispersal pattern, only the smallest-bodied primates can specialize on insects, although the benefits of supplemental insect consumption are seen across the order. Only tarsiers (*Tarsier* spp., body size ~120 grams) are obligate faunivores. They eat no plant matter and instead primarily consume a range of insects. They are also known to eat small vertebrates such as bats and birds. The slender loris (*Loris* spp., body size ~200 grams) is an insectivore; insects constitute over 50 percent of their diet. The rest of their diet consists of flowers, natural gums, and leaf shoots. Squirrel monkeys (*Samiri* spp., body size ~800 grams) are the largest primates that have a diet that consists of more than 50 percent insects during at least part of the year. The rest of their diet is mostly fruit, which they supplement with additional plant and animal resources. (Useful sources for quick comparisons of primate traits include Rowe 1996 and the National Primate Research Center's primate factsheets at http://pin.primate.wisc.edu/factsheets/.)

It can be quite difficult to quantify the insect consumption patterns of primates that are nonspecialist insect consumers. Primate species across the order ingest a lot of insects along with the leaves, fruit, or flowers the insects occupy. These insect foods are essentially invisible to onlooking researchers. Other behaviors are easy to observe, such as sifting through leaf litter on the ground to look for insects or visiting a colony of social insects such as ants or termites. Some of the most reliable data on eating insects comes from observations of great apes, who often go to great lengths to forage the nests of social insects.

The Great Apes

Our understanding of the great apes is indebted to three pioneering researchers: Jane Goodall, Dian Fossey, and Birute Galdikas. In the 1960s, these three were the first to observe and study great apes in their natural forest habitats. Famed paleoanthropologist Louis Leakey believed that the earliest hominins and their earlier ape ancestors would have used their environments in ways similar to those of extant great apes, but little was known about these secretive animals. Leakey believed that patience and perception were keys to making successful observations of how these animals lived. He also felt that women would be especially well suited to the task and that individuals not trained in anthropological theory would make less biased observations (Morell 1993). He sent Goodall, who was initially his secretary, to Gombe, Tanzania, to study chimpanzees. After Goodall's success in habituating chimpanzees and her reports of groundbreaking findings, Fossey and Galdikas approached Leakey to ask if they could follow her lead. Leakey sent Fossey to study gorillas and Galdikas to study orangutans (Morell 1993).

Both Goodall and Galdikas (Goodall 1963; Galdikas 1988) observed their focal subjects foraging for insects, but one of the most groundbreaking discoveries was Goodall's observations of chimpanzees fashioning tools to fish termites out of their nests. It had long been thought that tool making was an activity exclusive to humans, and when Goodall's report reached Leakey, he responded with the now-famous quote "now we must redefine tool, redefine Man, or accept chimpanzees as human" (Goodall 1998, 2184).

Although it sounds absurd to consider chimpanzees to be human, Leakey was not too far off. We now know from genetic evidence that the genus *Pan* is more closely related to the genus *Homo* than it is to the other great ape genera, even *Gorilla*, who often live in the same forests as chimpanzees (Chimpanzee Sequencing and Analysis Consortium 2005). Chimpanzee behavior is therefore used often in generating models of early hominin behavior.

Fossils represent once-living organisms, so it is understood that while those organisms were living they had to achieve the same basic requirements of life today: finding food, finding mates, and successfully reproducing. By studying chimpanzee behavior, researchers are able to identify how intelligent, large-bodied primates achieve these goals in their different environments. These findings are then used in paleoanthropology to create analogies for how early hominins, which were also intelligent, large-bodied primates, navigated their environments to achieve these same goals.

Although it is possible that tool use evolved independently in both chimpanzees and hominins, it is commonly assumed that their last common ancestor was intelligent enough to use simple tools (Wynn and McGrew 2001; Wrangham and Pilbeam 2001). This assumption reflects the principle of parsimony, which states that evolutionary explanations should not be more complicated than necessary and that when there are two competing hypotheses, the simplest one should have precedence (Futuyma 1998). Although the most parsimonious explanation may not always hold once more data become available, until then it offers researchers an opportunity to build scholarship even though they don't have all the answers.

Behavioral studies from ape species other than chimpanzees can also be useful in understanding early hominin life. For instance, a behavior such as obligate bipedalism, which is found early in hominin fossils but is not seen in chimpanzees or other apes, likely evolved independently on our lineage. However, a behavior that is seen across apes but is absent in humans today may still be a good candidate for early hominin reconstructions because it may be something that was lost later in our evolution. An example of this might be eating uncooked fibrous foods such as bark that are essentially impossible for the human gut to digest but are commonly consumed by nonhuman primates. With this is mind, studies of insect foraging by each of the great apes are helpful for understanding how these foods could have been used by our earliest hominin ancestors.

Chimpanzees

Termite fishing is a behavior seen at many chimpanzee study sites. In its simplest form, chimpanzees choose a twig, a vine, or a long blade of grass, clip off each end with their incisors, and poke it into the exit hole of a termite mound. Termite soldiers and workers respond to the breach in the nest, the soldiers by latching on to the tool with their mandibles. Once their mandibles close on their attacker, they do not release. The workers act to close the channel into the mound, blocking the soldiers from returning. The soldier caste is dispensable in this way to protect the rest of the colony. Chimpanzees use the latching mechanism of soldiers to their advantage, easily extracting the termites that are attached to their tool and consuming them with little risk of being bitten (figure 5.3). Variations exist in how chimpanzee populations across Africa use tools to fish for termites. Some fashion brush tips on the end of the tool by running a blade of grass through closed teeth, just posterior to their canines. A brush means that there is more surface area for the termite soldiers to bite, potentially increasing the payoff for each bout of fishing. Another variation is to use

Figure 5.3. Adult female chimpanzee Glitter fishing for termites (*Macrotermes subhyalinus*) at Gombe Stream National Park in Tanzania, November 2016. Chimpanzees at Gombe commonly use the hand that is not using a tool to support the weight of the termites as they eat directly from the tool. Photo by J. Lesnik.

two tools: first a stick to reopen passages the termites have sealed, then the traditional fishing probe (Sanz, Morgan, and Gulick 2004).

Chimpanzees are very selective about which termites they prey upon. Researchers have recorded chimpanzees consuming only eight genera of the eighty-five genera of termites present in tropical Africa (Bogart and Pruetz 2008; Lesnik 2014; Abe, Bignell, and Higashi 2000). Chimpanzees primarily feed on termites from the genus *Macrotermes*, often choosing them over more abundant termite species in the area. Collins and McGrew (1985) describe how chimpanzees at Mahale in Tanzania prefer *Macrotermes* over the widely available *Odontotermes*. They hypothesize that this is because the soldiers of *Macrotermes* are larger than the soldiers of other termite genera and because the exit holes in their nests are numerous and easy to find. Interestingly, rehabilitated chimpanzees released on Rubondo Island in Tanzania are known to include

Odontotermes in their diets (Moscovice et al. 2007). In fact, most of the genera of termites chimpanzees consume in rescue reserves are not the same as those natively wild chimpanzee groups select. At Rubondo Island, chimpanzees eat *Microtermes* and *Odontotermes*, both part of the Macrotermitinae subfamily of termites. These observations differ from those at the Ipassa Reserve in Gabon, where the genera chimpanzees consume include *Microcerotermes* and *Procubitermes* from the Termitinae subfamily and *Nasutitermes* from the Nasutitermitinae subfamily, even though *Macrotermes muelleri* are widely available (Hladik 1973). Chimpanzees also regularly consume *Pseudacanthotermes*, also of the Macrotermitinae subfamily, and *Cubitermes* of the Termitinae subfamily, but not as widely or in the quantities seen for *Macrotermes* (Uehara 1982; Newton-Fisher 1999).

Mound structure may be one factor that affects which termites chimpanzees choose to prey on. The basic architecture of a *Macrotermes* mound is a conical structure with entry and exit holes that tunnel into the many intercommunicating chambers. Most of the chambers in a *Macrotermes* nest contain fungus combs. *Macrotermes* maintain symbiotic relationships with this fungus: the termites bring in forage food and the fungus partially breaks it down before the termites consume it. This behavior requires that all food be brought back to the nest. Storage of food may correlate with a high density of termites in the nest and may be another reason why chimpanzees forage for termites from Macrotermitinae most commonly. Other foraging termites, like those of the subfamily Nasutitermitinae, do not always store their food or live in the same place as their stores. This disperses the population, which means a lower payoff for attacking their mounds (Sands 1965).

Macrotermes primarily use a mechanical defense strategy; they use their mandibular pincers to bite or pierce their enemies. Other termite genera use chemical defenses. Many species of termites have glandular devices that produce and deliver chemical weapons such as irritants, contact poisons, and glues (Prestwich 1984). Certain species of *Odontotermes* have hypertrophied salivary glands that contain quinines that can be used as nonspecific irritants. Glue spitting is the defense mechanism of termites of the subfamily Nasutitermitinae, which chimpanzees rarely consume. This viscous, sticky solution acts as a topical toxicant that can be deadly to invertebrate attackers such as ants. For larger predators, chemical defense mechanisms reduce palatability (Richardson 1987).

Some populations of chimpanzees live in forests that host all the same termites and all the same vegetation for tools that other chimpanzees use, yet they ignore the termites entirely. Because some research sites have been established for many years, the presence or absence of tool use is well documented, but not all of the variation can be readily explained by resource availability or other ecological factors (Whiten et al. 1999; Hedges and McGrew 2012). Primatologists have described these variations as cultural (McGrew 1992). To some anthropologists, the ascription of culture to a nonhuman primate is problematic because it is difficult to define culture, even for humans (Byrne et al. 2004). However, these sorts of non-ecological variations in behavior can be acquired only by being a member of a particular society, which fits the definition of culture offered by Tylor (1871). Although this definition is outdated when it comes to understanding human culture, for nonhuman primates, it helps identify behaviors that are acquired by means other than instinct alone.

It is well documented that adult female chimpanzees consume more insects than males. McGrew (1979) was the first to quantify this pattern from observations of extractive foraging and analysis of feces for insect remains. Lonsdorf (2005) has suggested that this pattern in adults corresponds to differences in the amount of time male and female juveniles spend learning about extractive termite foraging through observation. Lonsdorf found that young female chimpanzees started to fish for termites at a younger age than males, they were more successful than males once they had acquired the skill, and they used techniques that were similar to those of their mothers, whereas males did not (figure 5.4). Moreover, young females spent significantly more time watching other individuals perform the task. Lonsdorf noted that socioecological theory predicts that while female reproductive success is limited by access to resources, male reproductive success is limited by access to mates. Applied to chimpanzees, this theory predicts that adult females should primarily be interested in finding food and adult males should primarily be interested in establishing themselves socially. Social connections will help males gain access to females through the dominance hierarchy or through alliances. Thus, selection may shape which behaviors young chimpanzees are most adept at learning and thus which skills they carry into adulthood.

Chimpanzees also use extractive foraging techniques to prey on ants. A female bias for this technique has been observed, but we do not have corresponding data for the juveniles, as we do for the behavior of termite

Figure 5.4. A young female chimpanzee, 4.5-year-old Gossamer, modifying a piece of vegetation into a tool for termite fishing at Gombe Stream National Park in Tanzania, November 2016. Photo by J. Lesnik.

fishing. Chimpanzees primarily forage for army ants of the genus *Dorylus* by using long, stiff wands of woody or herbaceous vegetation to extract the ants from their underground nests. As is the case with termites, it is the soldier ants that attack the wand first, followed by the slower workers. These ants are highly aggressive; after the disturbance, increasingly more individuals join the defense as the colony works together to create an attacking horde. When chimpanzees forage for *Dorylus*, the payoff appears to be variable. The length of a wand, the aggressiveness of the ant species or colony, and mean ant size can all affect the rate of return (Schöning et al. 2008). An experimental study with *Dorylus rubellus* in Nigeria's Gashaka Gumti National Park showed that amassing ants provide a steadily growing harvest for the first seventeen minutes, after which the yield drops markedly (Allon, Pascual-Garrido, and Sommer 2012). Ant-dipping bouts are much shorter than termite fishing, which sometimes lasts for several hours. This difference may be because the yield of ant dipping decreases after the initial emergence of the soldiers or because of the increasing number of bites the ant-dipper gets from the ants since there is no way to restrict the ants to the wand (figure 5.5). Chimpanzees manage to secure some protection from bites by perching off the ground in a tree, but the ants will eventually overtake that vegetation. Because of the hazards of ant dipping, learning opportunities for juveniles are different from that for termite fishing. A study from Bossou, Guinea, found that mothers with young under five years old dipped more often at ant trails than at nests where the risk of an ant horde was less and opportunities for juveniles to observe and practice were greater (Humle, Snowdon, and Matsuzawa 2009). Although female chimpanzees have been observed ant dipping more than males at Mahale, Tanzania (Hiraiwa-Hasegawa 1989), sex differences were not observed at Bossou. The authors suggest this may be related to the fact that hunting for mammalian prey is rare at Bossou so males may be obtaining their protein from insects just as females commonly do (Humle, Snowdon, and Matsuzawa 2009).

Chimpanzees from the Goualougo Triangle in the Republic of the Congo are known to use several different types of tool sets when they forage for insects. They use a combination of long puncturing sticks and fishing probes to gather subterranean termites and a combination of short, stout perforating sticks and fishing tools to gather termites from their aboveground mounds. They also use several combinations of tool types to gather honey. Researchers have observed these chimpanzees using a tool set to prey on

Figure 5.5. Sequence of 6.5-year-old female chimpanzee Fadhila using a tool to dip for army ants (*Dorylus* spp.) at Gombe Stream National Park in Tanzania, August 2014. These aggressive ants can attack as a horde, so chimpanzees often perch themselves away from the nest for protection. Photo by Rob O'Malley.

army ants. Similar to the process they use for termite foraging, they first use a stick to perforate the nest of the army ant and then follow up with a herbaceous wand for dipping. Despite the similarities with termite foraging, these tools are unique to the task, distinguishable by both the materials used and by their smaller length and diameter. Termite-fishing tool sets are reported at several chimpanzee sites, but chimpanzees use tool sets less commonly for ant dipping (Sanz, Schöning, and Morgan 2010).

Chimpanzees also consume carpenter ants of the genus *Camponotus*, but despite the fact that these ants are broadly distributed, they appear to consume them at fewer sites than *Dorylus*. Although primatologists often consider these group-specific behaviors cultural variations (McGrew 1992; Whiten et al. 1999), there are few examples of successful cultural transmission between wild groups. It can be difficult to confirm cultural transmission, but one of the best-documented examples is for the spread of *Camponotus* ant fishing across different chimpanzee communities at Gombe National Park in Tanzania (O'Malley et al. 2012). At Gombe, there are three different communities of chimpanzees. The Kasekela community has been continuously studied since Jane Goodall began her work in 1960. The Mitumba community was habituated in the late 1980s, and the Kalande community has been monitored since 1999 but is not habituated. Researchers documented habitual fishing for *Camponotus* among the chimpanzees of the Mitumba community at Gombe soon after their habituation in 1992, but it was not regularly seen in the Kasekela community at that time. According to O'Malley and colleagues (2012), Goodall recorded potential observations of three chimpanzees fishing for *Camponotus* at Kasekela in 1978. These individuals were two immigrant females and an offspring of one of the females. By 2010, *Camponotus* fishing was customary at Kasekela, where, with one exception, it was practiced by immigrant females and by chimpanzees born in the group after 1981. Although using a tool to forage for *Camponotus* may have been an independent invention at Kasekela, these observations strongly suggest that an immigrant female brought the behavior, likely from the Mitumba community, and that observational learning has allowed the behavior to spread through the community.

Chimpanzees also exhibit marked sex differences regarding vertebrate prey. Males concentrate on obtaining mammalian prey, such as monkeys. They work together in groups to stalk, pursue, capture, and kill a single mammalian prey, then divide and distribute the carcass (Goodall 1963; Mitani and Watts 2001). This activity requires speed and agility and often occurs high in the tree branches, putting the males at risk of injury. Because males are unencumbered with dependent offspring, they are better able to engage in such risky activities.

Bonobos

Bonobos (*Pan paniscus*) belong to the same genus as chimpanzees (*Pan troglodytes*). These apes are understudied because the only area they live in, the Congo Basin in the Democratic Republic of the Congo, has been marked by civil war since the mid-1990s, and that has made it difficult for scientists to conduct research there. These apes are often viewed as the non-aggressive, "make-love-not-war" member of the *Pan* genus, for they rely on sexual activity to create social bonds and resolve conflicts. Their diets are more omnivorous than that of chimpanzees, and they are not known to use any foraging tools in the wild.

The insect portion of the bonobo diet is poorly understood, but unlike chimpanzees, they do not seem to focus on social insects. Most accounts of bonobo insect feeding are nonspecific and anecdotal (Badrian, Badrian, and Sussman 1981; White 1992), although consumption of millipedes, caterpillars, insect larvae, termites, and ants has been reported (Rafert and Vineberg 1997). McGrew and colleagues (2007) conducted the only systematic study of bonobos to specifically look at insect consumption. These researchers found potential insect prey at the site of Lui Kotale, but when they examined the fecal remains of bonobos, they did not find insect traces.

Orangutans

Orangutans, like chimpanzees, exhibit a female bias for using tools to extract insects, although their insect foraging is not documented at the same level of detail. Orangutans use tools to feed on termites, ants, and stingless bees that occupy holes in the trees, so it is harder to make observations of their behaviors than it is for the more terrestrial chimpanzees and gorillas.

Fox and colleagues (2004) provide the most thorough investigation of insect foraging by orangutans (*Pongo abelii*), a study of the population at the Suaq Balimbing Research Station in Sumatra, Indonesia. Females were shown to spend 18–24 percent more time feeding on insects than adult males and 53–73 percent more time than subadult males. Pregnant females spent nearly two hours per day feeding on insects, which was slightly but not significantly more than the time females with semi-dependent young spend. Lactating females spent the least amount of time foraging for insects, suggesting that being encumbered by their infants influenced the time they could spend with the task. This result is different from what is seen in chimpanzees, for whom reproductive state does not affect the rate of insect consumption (O'Malley et al. 2016). The disparity likely reflects the arboreality of the task for orangutans; mothers who are distracted by tool manipulation while high in the trees may increase the risk that dependent young will fall.

Although orangutans are generally considered to be solitary, data from the Suaq Balimbing Research Station suggest that individuals use tools at tree holes at different rates as a result of variations in opportunities to observe the techniques in other orangutans while growing up and perhaps later in life. It has been difficult to document social learning for orangutans, but data from experiments with captive orangutans seem to confirm field observations that orangutans pay attention to the goal of the task more than the methods of tool manipulation; social stimuli appear to inspire goal-directed practices in the play of juvenile orangutans (Call and Tomasello 1994).

Gorillas

Observing the rate of insect consumption is straightforward when tools are involved, but it can be more challenging to assess patterns for species such as gorillas that do not use tools. Deblauwe and colleagues (2003) found that 78 percent of fecal samples collected over a two-month period from western lowland gorillas (*Gorilla gorilla*) at Ntonga, Cameroon, contained insects. Ants were predominant, followed by termites, suggesting that social insects make up the bulk of the insect foods, but the researchers also identified the trace remains of Lepidoptera, Apoidea, Diptera and Orthoptera in the gorilla feces. Preferred termite prey included *Cubitermes* and *Thoracotermes*. The predominant ant prey belonged to *Oecophylla*, but *Crematogaster*, *Tetramorium*, and *Dorylus* were also common. There appeared to be significant variation in termite predation over the two months, but the ant portion of the diet varied only slightly.

Watts's (1989) study of *Dorylus* feeding by mountain gorillas (*Gorilla beringei*) at the Karisoke Research Center in Rwanda suggests that juveniles eat more ants than the adult females. To obtain the ants, the gorillas rushed to grab a handful from a colony or nest, then retreated quickly and frantically ate them from their hands. The dense hair on the backs of their hands provides some protection from bites, but it also traps the ants and the gorillas need to pick them out with their hands or lips. Often gorillas returned for another bout after this sequence. As the excavation of the nest proceeds, the gorillas may reach eggs and larvae, which they consume eagerly. Watts also reported that he never saw silverback males eating ants.

Another study of mountain gorillas from Bwindi Impenetrable National Park in Uganda found that females and juveniles ate *Dorylus* ants more than the silverbacks (Ganas and Robbins 2004). This study was conducted with fecal samples, but unlike the gorillas in the Ntonga study, these gorillas are habituated and researchers could determine the age and sex for the fecal samples they collected from night nests. Another example of female gorillas eating more insects than males comes from Mondika Research Center at the border of Central African Republic and the Republic of Congo, where researchers observed that female western gorillas ate more termites and ants (Doran-Sheehy et al. 2009). This study found an increase in insect eating during the rainy season that was likely due to the increased availability of this resource.

Not all studies that focus on the insect-eating behavior of gorillas found this female bias. A study of ant eating by eastern lowland gorillas from the Itebero region of Zaire (now the Democratic Republic of the Congo) found ants in fecal samples from all age and sex classes (Yamagiwa et al. 1991). A study of termite feeding by western lowland gorillas at Bai Hokou, Central African Republic, found that a higher proportion of fecal samples from the silverbacks contained the remains of termites than the samples from adult females and juveniles (Cipolletta et al. 2007). An interesting observation from this study is that altercations are common at termite mounds. When two individuals both want to feed at the same nest, the lower-ranking one will retreat. Termites are thus a sought-after resource for all individuals, but because of their dominance, silverback males get access more often. The gorillas at Bai Hokou most often targeted termites of the genus *Cubitermes*, but observers also saw them foraging at *Thoracotermes* mounds.

Tutin and Fernandez (1992) found that gorillas and chimpanzees living at Lopé Reserve in Gabon ate at least six insect species, only two of which were common to both groups. Both apes consumed *Oecophylla* ants most commonly. Each species exhibited different preferences for the other four prey insects: gorillas did not eat the insects that chimpanzees obtain with tools and chimpanzees avoided the small-bodied ants that gorillas go after. The authors suggest that these differences could complement the plant foods that predominate in the diets of each species.

Nutritional Contributions to Ape Diets

Deblauwe and Janssens (2008) investigated the insect prey preferences of chimpanzees and gorillas living at the same site at Dja, Cameroon. Their study, which provided detailed nutritional analyses, confirmed that the two ape species prefer different insect prey. The difference does not seem to be solely based on different access related to the use or non-use of tools. Nutritional analyses of the termite prey suggest that the insects these two groups select may supplement particular nutritional values in their diets. It is well documented that chimpanzees prefer *Macrotermes* soldiers, and in Cameroon, gorillas preferred workers of the soil-feeding *Cubitermes* termites. Because of these preferences, gorillas receive more iron and other micronutrients while chimpanzees receive more protein from their termite resources. Gorillas do not need to supplement protein in their diets because the herbaceous plants they consume are high in protein; however, they may not eat enough fruit to fulfill their micronutrient requirements. In contrast, chimpanzees eat plenty of fruit and receive enough micronutrients, but fruit specialists have a more difficult time obtaining protein from the foods they eat. Chimpanzees and gorillas have shared habitats over the course of their evolution, so their success relies on their occupation of different niches, including their use of different insect resources. Deblauwe and Janssens note that *Oecophylla* ants are used in Chinese and native Australian medicine to treat various kinds of illnesses, including gastrointestinal disorders, fevers, and inflammation. Extrapolating from this observation, the authors speculate that gorilla insect preferences may give them some relief from digestive upset related to their leaf-based diets, noting that such diets are often high in tannins and other plant toxins.

Little is known about which insect species orangutans exploit or the nutritional value of the insects they eat. Orangutans are predominantly frugivorous, but there are times when fruit is less available and does not make up the majority of their diet. Fox and colleagues (2004) found that insect feeding did not differ significantly with the availability of fruit. Insects thus constitute a valued part of the orangutan diet year-round and are likely important sources of protein, especially for females.

Bonobos do not regularly feed on insects, but they may not need significant supplementation since their omnivorous diet is rich in both fruit and terrestrial herbaceous plants. Additionally, the female-based social system and lack of aggression in bonobo societies means that females are less stressed and have full access to available resources, including meat. For these reasons, reproductive bonobo females may not need the additional supplements to their diets, as other ape females do.

Other Non-Insectivorous Primates

Insectivory patterns are more difficult to assess in smaller-bodied primates such as Old World and New World monkeys. Many of the insects these primates incorporate into their diets are embedded in the leaves and fruit on which they regularly feed. Many accounts of dietary breadth group all invertebrates together. For my research, this is problematic since different insects may offer different nutrition. Additionally, many studies do not break down data by sex and age classes when discussing the foods a particular species consumes.

There is some indication that female Old World monkeys show a preference for insect feeding. This warrants further investigation. McGraw and colleagues (2011) found no difference in the amount of invertebrates male and female sooty mangabeys (*Cercocebus atys*) consumed, but females spent significantly more time chewing the invertebrates they ate. Although this finding does not yield a clear interpretation, it suggests a difference in the use of the resource and possibly suggests that females give insects more attention. Isbell (1998) identified arthropods as the fourth most numerous food group for adult females and juvenile savanna-dwelling patas monkeys (*Erythrocebus patas*), behind swollen thorns, gum, and flowers. For males, the fourth most numerous food item was seeds, behind gum, swollen thorns, and flowers.

The insectivory patterns of New World monkeys are better documented. Insect prey makes up a large portion of the diet of both male and female capuchin monkeys. However, Melin and colleagues (2014) detail significant differences in insect choice between male and female white-faced capuchin monkeys. Females were more efficient at capturing soft, sedentary insects, primarily caterpillars, and spent more time feeding on colonizing insects such as ants and termites. The authors note that these results support previous reports of females who specialize on reliable prey sources, but they may also reflect the females' preferences for softer food types that are easier for individuals with smaller teeth and jaw muscles to consume. It may also be worth investigating the dietary contributions of these insects, since softer foods often represent larval stages, such as grubs and caterpillars, which tend to have a high fat content and caloric density.

Female squirrel monkeys (*Samiri oerstedi*) eat significantly more arthropods than males in all seasons except the wet season, which is the time of greatest food scarcity (Boinski 1988). Montague and colleagues (2014) confirm that female squirrel monkeys of *Samiri sciureus* also are more successful insect foragers than males and relate their higher capture rates to the close proximity of other females in foraging groups. Having neighbors nearby helps thwart the evasive behaviors of the insect prey, which may jump or fly to escape one monkey but end up in easy reach of another. Genetic analyses revealed that female squirrel monkeys tend to stay near their kin, thus foraging success is a family affair. The advantage of having family nearby may explain why females of this species do not disperse from their natal group.

There are numerous informal accounts of female monkeys searching for and consuming insect prey more often than the males. Garber (1987) lists ten primate species in addition to chimpanzees for which the females devote more of their foraging effort to acquiring higher-protein resources such as leaves and insect prey; the males of these species spend more time consuming fruit. These species include both New World and Old World monkeys and one strepsirrhine. Future research projects dedicated to the insect portion of the diet of these species may reveal more distinct patterns that would be useful for modeling the benefits of insect foraging for female primates.

* * *

Humans are members of the order Primates. It can be difficult to see past our modern ways of life and recognize ourselves as such, but traits like our forward-facing eyes, grasping hands, and offspring dependence remind us of our origins. It is also important to remember that most of our modern developments, such as agriculture, civilization, and industry, have emerged within the last 10,000 years of human history. Before this time, our similarities with nonhuman primates, especially chimpanzees with whom we share over 95 percent of our DNA (Chimpanzee Sequencing and Analysis Consortium 2005), would have been more striking. For these reasons, behavioral data collected by primatologists, including their observations on insect consumption, are critical to our understanding of the lives of earlier hominins.

Among non-insectivorous primates that include insects in their diets, females tend to eat more insects than their male counterparts. This pattern is especially well documented for chimpanzees and orangutans who use tools to forage for insects. When females have young offspring by their sides, the juveniles, both male and female, acquire foraging techniques by observational learning and benefit from eating insects during this critical time in their development. As these primates mature, it is the females who continue to use tools to harvest insects as adults. This likely helps them meet their nutritional requirements during reproduction. Even in species that do not use tools, there are notable accounts of female preferences for insects. The pattern of a female bias for insect consumption is reported only when insect foraging and sex differences are the primary goals of the research. The data from larger surveys of foods consumed at sites often do not include these variables; thus, there is still much to learn about this pattern across primate species.

The currently available data show that the female preference for insects exists for many taxa of the Haplorrhine suborder. Fossil evidence suggests that the earliest members of this suborder were primarily insectivorous, similar to tarsiers (Ni et al. 2013), and that the earliest anthropoids (ancestors of non-tarsier Haplorrhines) occupied a different niche by being insectivore-frugivores (Kay, Ross, and Williams 1997). Any time since this initial dietary diversification, female members of Haplorrhine species may have found it beneficial to rely more on insects than their male counterparts. The pattern we see today does not necessarily have to have a single origin in Haplorrhine evolution; it could have evolved independently countless times. It is difficult to say which hypothesis represents the simplest explanation, but either way, it suggests that this behavior is a successful one for female primates to employ.

When we apply this to our understanding of how women in foraging societies target edible insects, it appears they are utilizing a validated primate strategy. Thus, the conflict model rather than the cooperative provisioning model is most useful for understanding how evolution contributed to the sexual division of labor for this resource. The divergence of the hominin lineage from our ape ancestors marks the beginning of the evolutionary trajectory that led to the gendered foraging patterns we see in modern hunter-gatherers. This path includes the evolution of larger relative brain sizes, increased offspring dependency, and the technological and social advances necessary to support these changes.

6

Reconstructing the Role of Insects in the Diets of Early Hominins

When it was time to start thinking about the direction my research should take, all I knew was that I had a special liking for early hominins and that I preferred studying behavior over morphology. But that was about it—how to pull those two things together in a research project was a mystery to me. I spent that summer in the Ditsong Museum of Natural History (called the Transvaal Museum at the time) in Pretoria, South Africa, and collaborated with Director Francis Thackeray on small projects as I thought about my thesis (Lesnik and Thackeray 2006; Lesnik and Thackeray 2007). While there, I got to see all of the fossils and artifacts the museum had from famous sites in the Cradle of Humankind: Sterkfontein, Swartkrans, and Kromdraai. Because of my interest in behavior, the tools from these sites especially drew my attention. I was encouraged to look at the bone tools from the Swartkrans site, but I was hesitant because it was not that long before then that Lucinda Backwell and Francesco d'Errico had published extensively on them, reporting evidence of their use to dig into termite mounds (Backwell and d'Errico 2001; d'Errico and Backwell 2003; Backwell and d'Errico 2005). What could I possibly contribute? It took some time, but I ultimately realized that these bone tools had the potential to change how we think about hominin diets. Backwell and d'Errico's research provided the evidence that these bone fragments were used to procure termites. With their findings, I could reconstruct the behavioral ecology of termite consumption in a way that had never been done before.

Paleoanthropologists have a thin understanding of termites as a food source. They generally agree that termites are nutritious, but they overlook their variability. Eighty-five genera and about 1,000 species of termites exist in tropical Africa (Abe, Bignell, and Higashi 2000).

Figure 6.1. This commonly used depiction of human evolution is known as the March of Progress. The diagram is a gross oversimplification and does not accurately demonstrate how the pictured species are related to each other.

The nutritional contributions of termites vary at the species level and even at the caste level (Lesnik 2014). After the discovery of evidence for hominin termite foraging, the next step was to determine which termite species they were going after. With this information, the nutritional offerings of termites can be more confidently included in reconstructions of hominin diet. In the previous chapters I identified preferences and means for insect consumption among both human and nonhuman primates. In this chapter and the next I apply these models to human ancestors.

The Hominin Lineage

About five or six million years ago, the evolutionary lineage that leads to humans diverged from the one that leads to the genus *Pan*, which includes present-day chimpanzees and bonobos. All of the organisms that evolved along our lineage since that point, including both our direct ancestors and dead-end offshoots, are collectively known as hominins. Outside the field of anthropology, the only representation of the hominin lineage that people are likely familiar with is the so-called March of Progress (figure 6.1). This image is a gross oversimplification of human evolution because evolution does not happen in a linear fashion. Also problematic is the fact that the first figure is an image of a chimpanzee, a living relative of ours, not an ancestor.

Figure 6.2. Modification of the March of Progress diagram that demonstrates the relationships among the five figures. Chimpanzee and modern humans are extant species and occupy the highest level of the diagram, while the extinct hominins branch from the lineage that led toward humans. The star represents the last common ancestor of humans and chimpanzees, which is estimated to have lived five to six million years ago.

In a modification of this diagram (figure 6.2), living species, humans and chimpanzees, are at the top, and the branch-like pattern represents how closely each of the extinct species is related to the ones that still survive. These silhouettes represent a small sample of the taxonomic diversity of the hominin fossil record. Key members of the human family tree are depicted in figure 6.3, although even this is not exhaustive and our understanding of the relationships between species is constantly changing as new evidence is discovered.

It could be argued that the image of the chimpanzee in figure 6.1 is meant to represent the last common ancestor of humans and chimpanzees, because until recently, the last common ancestor was thought to be a knuckle-walker like the nonhuman great apes. Fossil evidence for this important time in our evolution has been scant, but in 2009, the description of the fossilized bones of a rather complete skeleton of the hominin

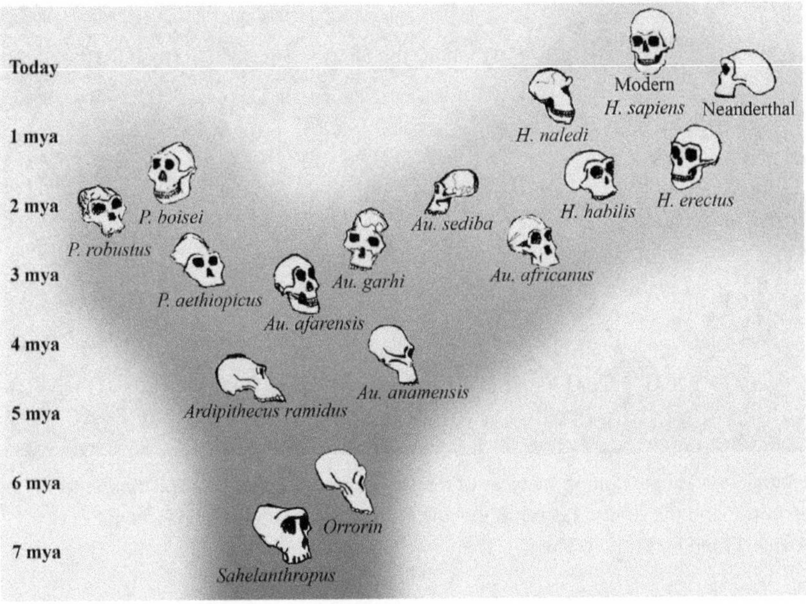

Figure 6.3. Estimated ages of and relationships among key fossil hominin species. Artwork by John Anderson.

Ardipithecus ramidus was published (White et al. 2009). These fossils are dated to about 4.4 million years ago, which makes them too young to represent the node between humans and chimpanzees. However, they shed light on what the last common ancestor probably looked like. One key piece of information this skeleton provided is that the last common ancestor was likely a smaller-bodied, above-branch clamberer. From this ancestral condition, both the hominin lineage and the panin (chimpanzees and bonobos) lineage adapted different styles of locomotion as body size increased. The standard depiction in the March of Progress perpetuates the misconception that "we came from chimps" and distorts the fundamentals of evolution and speciation. Each of the currently living species has undergone millions of years of independent evolution, acquiring many traits that are different from what was present in the last common ancestor.

The second figure in the March of Progress is meant to represent the australopithecines, the group this chapter focuses on. The word "australopithecine" is a collective term for the multiple species within the genus *Australopithecus*, which lived in Africa from 4.2 to 1.5 million years ago. The term is also used for species closely related to this genus whose

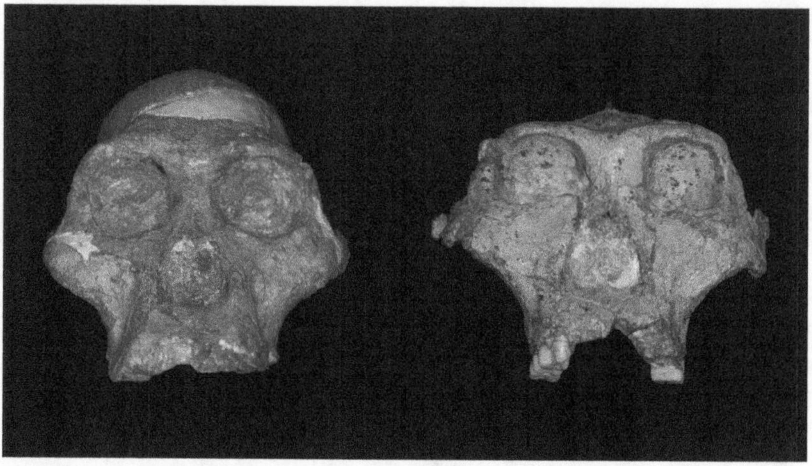

Figure 6.4. Comparison of the faces of the gracile australopithecine *Australopithecus africanus* and the robust australopithecine *Paranthropus robustus*, both from South Africa. Photo by M. H. Wolpoff.

inclusion within it is debated amongst paleoanthropologists. The earliest australopithecines were obligate bipeds. This feature enabled paleoanthropologists to place them on the hominin lineage with confidence. Other early australopithecine features represented a mix of ape-like and human-like characteristics. The small brains and projecting faces of australopithecines made their crania seem ape-like at first glance, but their teeth had distinctively human traits. Their canine teeth were reduced in size and were not sexually dimorphic and the molars had a thick layer of enamel. The molars resemble the teeth of humans today, not the teeth of our thinner-enameled ape cousins. The fact that both frugivorous chimpanzees and folivorous gorillas have thin enamel suggests that thicker enamel is adapted to more variable diets that include larger amounts of tough foods.

Australopithecines are commonly divided into two groups: gracile and robust (figure 6.4). Robust australopithecines are united by a suite of features on their crania that indicate strong muscle attachments and powerful jaws. These hominins would have been able to eat very tough foods and likely did during at least part of the year. Because of these distinct traits, robust australopithecines are often placed in their own genus, known as *Paranthropus*, while the gracile australopithecines maintain the moniker *Australopithecus*. Both gracile and robust australopithecines are found in both eastern and southern Africa, the two regions where hominin fossils are most abundant. Hominins likely ranged across the entire continent, but the environment for preservation and recovery is most favorable in these regions. The site of Swartkrans in South Africa yielded the largest number of fossils ascribed to the robust australopithecine *Paranthropus robustus.* This site also yielded a large number of the bone tools that Backwell and d'Errico (2001) identified as digging implements for termite nests. Thus, *P. robustus* is central to the model I have created in this chapter.

The remaining hominins in figures 6.1 and 6.2 represent *Homo erectus*, Neanderthals, and modern humans, the subjects of the following chapter.

Australopithecine Behavior and Tool Use

Behavioral reconstructions for hominins heavily rely on the preservation of tools in the archaeological record. Unfortunately, perishable tools like those chimpanzees and orangutans use to collect social insects did not preserve. The last common ancestor could have used perishable tools in this way, as edible insects would have been available across Africa back then, as they are now. Termites of the genus *Macrotermes* evolved in Africa about 20 million years ago. They are relatively unchanged today, even though dramatic environmental shifts have occurred since they evolved (Brandl et al. 2007).

The fossil and archaeological records of the australopithecines are abundant enough to support robust interpretations of behavior. The earliest reconstructions of australopithecine behavior portrayed them as fierce hunters. Raymond Dart (1925), who identified and named the first *Australopithecus* fossil in South Africa, believed that these hominins developed a material culture he called osteodontokeratic (Dart 1957), whereby they used tools made of bone, tooth, or horn to hunt large animals and violently fight each other. This hypothesis was met with contention and inspired seminal research in the area of taphonomy, the natural processes that affect bones between death, fossilization, and recovery (Brain 1981). The work of C. K. Brain (1981) demonstrated that the bone fragments found at hominin sites show the same pattern of preservation as bones found in modern carnivore dens. Certain parts of the body that have little meat on them, such as the hands and feet, tend to be preserved undamaged while the marrow-containing long bones are almost always found fragmented. The implements Dart identified as tools are better understood as natural bone fragments that resulted from carnivore kills. Since the remains of South African australopithecines fossils are found in the same fragmentary pattern as the other animal remains at the sites, these hominins may be better portrayed as the hunted rather than the hunters (Brain 1981; Lee-Thorp, Thackeray, and van der Merwe 2000).

Even though Dart's theory of an osteodontokeratic culture has been refuted, we know that australopithecines were notable tool users. The earliest evidence of modified stone tools is nearly 150 artifacts found at the site of Lomekwi on the western side of Lake Turkana in Kenya that dates to 3.3 million years ago (Harmand et al. 2015). These tools were created with a technology similar to the core-and-flake technology known as the Oldowan industry, although the discoverers suggest that there are distinct differences in the assemblages and propose the name Lomekwian for the older tools (Harmand et al. 2015). The tools in both groups were formed using the same methods. A percussion tool, or hammer stone, is used to strike a sharp flake from a second stone known as the core. The removal of flakes from cores leaves distinct patterns on stone that are different from what occurs through natural fissuring. This is what enables paleoanthropologists to identify these stones as artifacts in the archaeological record.

One of the best sites for evidence of early tool use is the 2.6-million-year-old site of Gona, Ethiopia, where Oldowan tools were found in context with hominin fossils ascribed to the species *Australopithecus garhi*. Animal bones that contained cut marks that are evidence of butchery using stone tools were also found at this site (Semaw 2000; Dominguez-Rodrigo et al. 2005). The Gona site is extraordinary because it has well-preserved cut marks in the same context as the tools and the hominins. Possibly more interesting is that the cranial capacity for *Au. garhi* is estimated to be 450 cubic centimeters (Asfaw et al. 1999), very similar to that of chimpanzees, suggesting that the increase in australopithecine brain size occurred after the advent of these tools, although important neural reorganization may have occurred by this time (Holloway, Broadfield, and Yuan 2004).

The evidence of butchery found at Gona seems to be the exception rather than the rule for early australopithecine sites. Widespread evidence of butchering does not appear until much later in the archaeological record and is generally associated with members of the genus *Homo*. The largest-brained australopithecines are the robust australopithecines, whose brain sizes increased by about 20 percent to 500–550 cubic centimeters. These hominins, which lived about 1.7 million years ago, were contemporaneous with the earliest members of the genus *Homo* and are often found at the same sites (Leakey 1971; Brain 1993). When tools and evidence of butchering are found at sites that contain members of both genera it is difficult to determine which hominin was responsible.

For the genus *Homo*, large brains were supported by high-quality diets that included increased amounts of meat and bone marrow and tool-assisted food processing (Milton 1999a; Navarrete, van Shaik, and Isler 2011). The archaeological record for australopithecines suggests that meat consumption was more rare for these hominins but the increase in their brain size over time suggests that some energetic tradeoff must have occurred. According to one theory, an increased breadth of the australopithecine diet to include resources that could be acquired by simple tools such as digging sticks may be the defining feature of the hominin line (Mann 1972). This hypothesis never gained much traction since tools made of wood or any other perishable material did not preserve in the early archaeological record. However, it is worth revisiting this idea in light of the South African bone tools.

The bone artifacts found at the *P. robustus* sites of Swartkrans and Drimolen in South Africa are a departure from tool assemblages found at numerous hominin sites (Backwell and d'Errico 2008) (figure 6.5).[1] These bone tools are fragments of animal long bones. The rounded and polished surface of one end suggests that they were digging implements. There are over 100 such artifacts at these two sites. Taphonomic explanations cannot explain how polish could be found only at one end of a bone fragment in numerous instances across different contexts. Thus, the evidence is convincing that these are digging implements that the hominins used (Backwell and d'Errico 2001).

The surface wear on the ends of these tools has been studied with the goal of identifying the particular task for the tools. These studies used different methods, ranging from qualitative analysis of scanning electron microscope images to quantitative measuring of wear features from light microscope images to three-dimensional renderings of roughness features and surface texture (Brain and Shipman 1993; Backwell and d'Errico 2001; van Ryneveld 2003; d'Errico and Backwell 2009; Lesnik 2011). This body of literature suggests that the wear patterns on the ends of the bone artifacts are variable and do not fit entirely within the range of variation of any one particular task. However, the theory that they were primarily

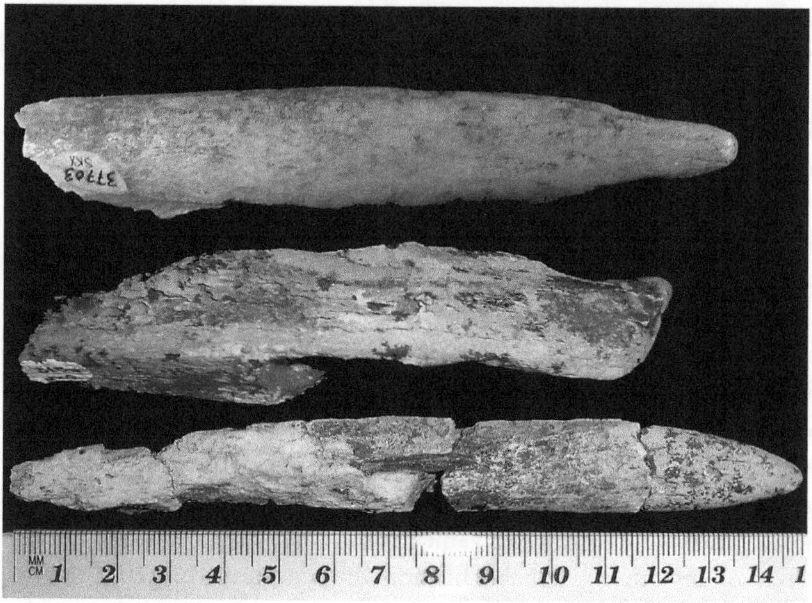

Figure 6.5. Bone tool artifacts from the site of Swartkrans in South Africa. Photo by J. Lesnik.

used for digging into termite mounds is supported in multiple follow-up studies (van Ryneveld 2003; d'Errico and Backwell 2009; Lesnik 2011). Researchers have found that when experimental tools are used to excavate termite mounds, the result is striations that are uniformly parallel to the long axis of the tool and narrow in width. These marks are consistent with digging into the fine sediments of termite nests. Digging into the ground for resources such as tubers or bulbs creates less uniform striations in both size and direction, since rock inclusions are present in the soil and the task requires horizontal movements to dig out the fleshy part of the plant.

Using a bone tool to dig into a termite mound would have created two opportunities for australopithecines that generally are not available to nonhuman apes today. First, excavating the top layer of a termite mound can reveal more passages, since termite workers conceal their presence at the surface. These protected passages often contain more termite activity. This is the technique that I use when I collect samples of termites for chemical analyses. Second, the use of a digging tool suggests that termite nests could not easily be opened by hand, unlike the mounds of

Figure 6.6. A woman using a broom made from nearby vegetation to collect large numbers of *Macrotermes* termite soldiers in Limpopo, South Africa. Photo by J. Lesnik.

Cubitermes, which are easily toppled. Gorillas are able to access the central chambers in *Cubitermes* mounds where the nymphs and larvae reside, but chimpanzees who fish at *Macrotermes* mounds cannot get to them. Using a tool, the australopithecines could access these inside chambers of hard-shelled mounds and collect the fat-rich termite young within. Additionally, when a mound is opened completely, larger and more efficient tools such as brooms can be used to collect termite soldiers in larger numbers (figure 6.6).

Australopithecine Diet

Termites may have supplemented the diet of australopithecines, such as *P. robustus* at Swartkrans. Chimpanzees supplement their frugivorous diets with protein-rich *Macrotermes* soldiers and gorillas supplement their folivorous diets with micronutrient-rich *Cubitermes* workers. Humans consume at least sixty-one species of termites, most of which belong to the genus *Macrotermes*. However, unlike chimpanzees, people tend to prefer the larvae and alates to the soldiers.[2] These castes have a higher fat content than the other castes in their species. In order to determine which termite prey would have been the most desirable to hominins, their general diet must first be reconstructed to determine what supplemental nutrients were necessary. When we have this information, we can assess whether the protein-rich chimpanzee model, the micronutrient-rich gorilla model, or the more-varied-with-emphasis-on-fat human model would be the best fit for reconstructing the termite portion of the australopithecine diet.

Stable isotope analysis has become the standard method for analyzing ancient diets. This research is based, quite literally, on the concept that you are what you eat. Our bodies take in different forms of elements known as isotopes from the foods we eat and store them in our bones and teeth. Some of these isotopic signatures preserve in the chemistry of dental enamel for millions of years, giving us a peek at what hominins were eating. These signatures yield only general evidence of some of the foods that were consumed, but this approach has been particularly important in rethinking the diet of australopithecines.

One of the most useful elements in dietary stable isotope analyses is carbon. Plants have two distinct ways of storing carbon, and since plants are at the bottom of most food chains, they contribute these signatures to the entire ecological system they support. These two pathways put plants into categories called C_3 and C_4, each of which stores different ratios of the stable isotopes ^{12}C and ^{13}C, known as $\delta^{13}C$ and read as "delta 13 C." C_3 plants tend to be woody, and animals that eat any component of that plant—bark, leaves, fruit, and so forth—will have a C_3 signature in their dental enamel. C_4 plants, which are typically grasses, are specialized to use an additional photosynthetic pathway. These plants store a larger amount of the heavier ^{13}C isotope and create a distinct $\delta^{13}C$ signature. This C_4

signature is present in the animals that eat the plants and is even preserved at the higher trophic levels in predators that hunt grazing herbivores.

Beginning in the mid-1990s stable carbon isotope analysis of tooth enamel was used to study the diets of early hominins from South Africa (Lee-Thorp, van der Merwe, and Brain 1994; Sponheimer and Lee-Thorp 1999; van der Merwe et al. 2003; Sponheimer et al. 2005). Isotopic evidence of the diet of South African hominins from Swartkrans suggests that the diet consisted of 60–65 percent C_3 foods and 35–40 percent C_4 foods (Sponheimer et al. 2006). Therefore, *P. robustus* did not have an exclusively frugivorous diet and instead consumed significant amounts of C_4 foods. This finding was unexpected. Before the application of isotopic research to hominin fossils, dental evidence for the diet of australopithecines suggested their diets were dominated by fruits and seeds (Kay 1985).

It may seem that these isotope results should initiate a reinterpretation of the sites with evidence of butchering where robust australopithecines and the genus *Homo* coexisted, but earlier South African hominins attributed to the species *Au. africanus*, which share the same $\delta^{13}C$ signature, are not known to even use tools (Sponheimer and Lee-Thorp 1999). In the absence of tool use, it is difficult to construct a dietary model that included eating enough meat to make up 35 percent of the diet for these early hominins. C_4 grasses, C_4 sedges, C_4 forbs (herbaceous dicots) were foods suggested in addition to meat as possible contributors to this signal, but it became clear that there was a lack of data available on carbon isotope composition of other potential C_4 foods (Peters and Vogel 2005; Sponheimer et al. 2005).

At roughly the same time as these isotope studies were published, Backwell and d'Errico (2001) published their work on the Swartkrans bone tools. The authors made their own experimental bone tools and used them to dig into the ground for tubers, to strip bark off trees, to tan hides, to process marula fruit, and to dig into termite mounds of the genus *Trinervitermes*. Using a variety of microscope techniques, they compared the wear patterns on the ends of the experimental tools to the wear patterns preserved on the ends of the Swartkrans artifacts. They found that the best match was the tools used to dig into the mounds of locally abundant *Trinervitermes* termites. Because *Trinervitermes* is a grass-foraging genus, these results suggest that if the hominins were eating these termites, they may have contributed to the C_4 results found in the $\delta^{13}C$ signature.

The exciting idea that termites could provide the answer to the conundrum of the isotope signature of australopithecines is how I became interested in the topic of insects as food. I began to familiarize myself with termite diversity and ecology and had the realization that referring to termites as a food source was essentially meaningless if we could not identify which exact termites were being consumed. The termites in the Backwell and d'Errico study happened to be grass foragers, but termites are most commonly associated with eating wood. The type of termites the hominins ate would have determined what they contributed to hominin nutrition. Additionally, termites are variable within the species level because they are social insects with a caste system. Termite soldiers provide an entirely different nutritional resource than the termite nymphs in the same nest. In order to seriously consider termites as part of hominin diet reconstructions, we need to be clear about which termites. As appealing as *Trinervitermes* are as candidates for a C_4 resource, there is not much evidence that they make an attractive food option. Available reports show that people consume only twelve of the eighty-five genera of termites found in sub-Saharan Africa, and *Trinervitermes* is currently not on that list of twelve (Jongema 2017). Additionally, none of the extant great apes eat *Trinervitermes*. These termites have a mechanism by which they emit a toxicant that defends them against ants and other predators. Even aardvarks, which regularly feed on termites, eat *Trinervitermes* only for a short while because of this chemical deterrent (Taylor, Lindsey, and Skinner 2002). I began to look at other termite genera as possible sources of hominin food.

My original project involved more experiments with digging into the mounds of termites of *Trinervitermes* and *Macrotermes* using bone tools (Lesnik 2011). These two genera build distinctly different mounds. *Macrotermes* produce mounds that often reach meters into the air and have a concrete-like outer crust. *Trinervitermes* create low, rounded mounds that reach less than half a meter in height. The crust is firm but not too tough and the inner matrix is soft. I believed that these strikingly different structures would produce different wear patterns on the bone tools I was using to breach the crust of a mound and dig inside it. However, my results were not so clear. I cannot definitely say which genus of termites hominins were using bone tools to forage for. My results suggest that the evidence we have for termites in the diet is nonspecific and needs refinement. I thus worked to create a model of the diet of *P. robustus* in order to assess which genus of termites made the most sense nutritionally as prey.

To start with, the isotope conundrum may be less of a conundrum than originally thought. Additional evidence from another robust australopithecine, *P. boisei* from Tanzania, suggests that the C_4 component of their diet was even greater than for South African hominins. This means that marginal resources, such as insects, are less likely to have been responsible for such a large component of the hominin's diet (Cerling et al. 2011). More widely available resources thus provide a more compelling answer to the C_4 conundrum. These resources might include grasses and sedges, including the underground storage organs of these plants, but they also may include a different type of plant known as CAM plants. CAM plants are plants like cacti that are adapted to arid environments and have their own photosynthesis process that has an isotopic signature that is similar to that of C_4 plants. In my model, I use the nutritional values of grasses, sedges, and ferns (which are often CAM photosynthesizers) to reconstruct the C_4 portion of the diet. The C_3 component of the hominin diet is likely dominated by fruit and leaves, just as in extant apes. If we assume that the diet of hominins likely improved in quality during their evolution, there is no reason to hypothesize that fruit consumption decreased when grasses and sedges were added to the diet.

The habitat of the South African australopithecines was likely not dense forest (Vrba 1975). This means that it was likely different than the habitat we see for chimpanzees in equatorial Africa today. However, this probably did not affect the availability of fruit as a food resource. Observers of chimpanzees feeding in dry savanna environments suggest that they find similar amounts of fruit as chimpanzees in equatorial Africa; they spend roughly 62 percent of their feeding time consuming fruits (Pruetz 2006). Thus, it is possible that hominins in an open woodland-savanna environment also consumed significant amounts of succulent fruit. It can be conservatively assumed that hominins were able to find enough fruit to account for half of the C_3 portion of their diet. This would have meant that roughly 30 percent of their total diet came from fruit (figure 6.7). The remaining portion of C_3 resources would likely have consisted of other woody plant resources, most commonly foliage. Supplementary resources such as invertebrate and vertebrate prey, consumed opportunistically, would have contributed to both signatures.

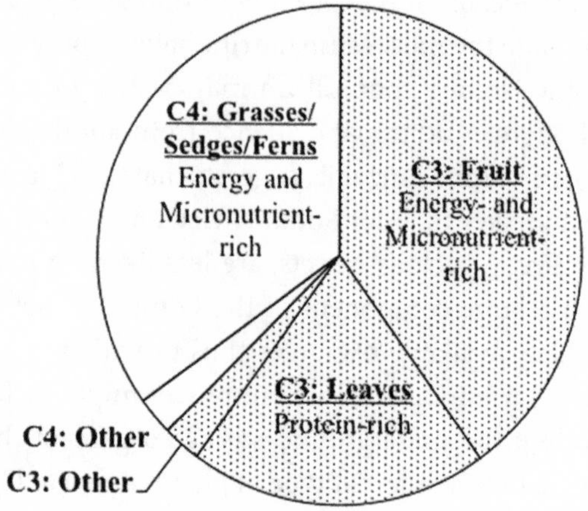

Figure 6.7. Proportion of foods in early hominin diet as modeled for *P. robustus*. Protein-rich foods appear to have made up only a small portion of the average diet.

An analysis of the nutrient composition of major items chimpanzees consume shows that leaves contain over twice as much protein as fruit and more fat than fruit (see table 6.1). The nutrient composition of pith, the spongy inside of vascular plants and a common chimpanzee fallback food, resembles fruit more than leaves and thus is an appropriate substitute for fruit when its availability is low (Wrangham et al. 1991). An assessment of savanna plant foods in Kruger National Park in South Africa shows that woody plants consistently have more protein than grasses (Codron et al. 2007). The crude fat in four species of savanna grasses present in Sudan averages 1.4 percent of dry matter, lower than the values seen in leaves and fruit but similar to the values seen in pith (Savadago et al. 2009). Considerable amounts of phosphorous, calcium, magnesium, and potassium can also be obtained from savanna grasses, especially those found under tree canopies (Ludwig, Kroon, and Prins 2008). Thus, grasses in the hominin diet likely provided carbohydrates and minerals but not fat or protein.

Sedges, like grasses, are rich in energy but low in protein. *Cyperus papyrus*, a sedge that is available year-round in the floodplain grasslands of southern Africa, produces raw culms and fleshy rhizomes that are high in carbohydrates and gross energy but low in fat and protein; they yield less than a half a gram of fat and up to one gram of protein per 100 wet grams (van der Merwe, Masao, and Bamford 2008). Ferns are similarly low in

Table 6.1. Estimated nutritional content of major groups of hominin plant foods

	Crude Protein (% DM)[1]	Crude Fat (% DM)	Micronutrients
Fruit	9.5	4.9	high
Leaves	22.1	1.4	low
Grasses, underground storage organs	10.5	1.6	high
Ferns	10.2	9.5	high

1. DM = dry matter.

Sources: Values for fruit and leaves obtained from Conklin-Brittain, Wrangham, and Hunt (1998); values for grasses and underground storage organs based on Codron et al. (2007); values for ferns obtained from Bassey et al. (2001).

protein, high in energy, and high in micronutrients, but unlike sedges and grasses, they contain more fat. A particular edible fern that the peoples of the Akwa Ibom State of southern Nigeria consider a delicacy contains about 10 percent protein, similar to grasses and sedges, but also 10 percent fat, well above the average 2.6 percent fat found in leaves (Bassey et al. 2001).

Using these nutritional values for probable hominin foods, if the hominin diet was about 30 percent fruit, about 30 percent leaves and other fibrous tree products and about 40 percent grasses, sedges, and ferns, then it appears that protein deficiency could have been a concern for *P. robustus* and other australopithecines. Although it is possible for chimpanzees to meet their protein requirements by consuming large quantities of low-protein fruit, they are known to supplement this diet with various protein-rich sources such as seeds, nuts, and both vertebrate prey and invertebrate prey (Milton 1999b). Because we know that robust australopithecines had low C_3 signatures, we can assume that they were not consuming leaves or fruit at the levels of extant apes, so protein supplementation may have been even more important in their diets than it is for apes today.

Hominins could have benefited from consuming protein-rich *Macrotermes* soldiers in the same amounts as chimpanzees today. It is difficult to estimate how many termites a chimpanzee manages to collect when visiting a termite nest, but one estimate from Dja, Cameroon, suggests that the daily rate of insect consumption is 14 grams of dry matter. This estimate, which included both ants and termites, would yield almost 12 grams of protein (Deblauwe and Janssens 2008).

It is estimated that chimpanzees require 0.59 grams of protein per kilogram of body weight each day (National Research Council 2003). The average weight of male chimpanzees is 50.6 kilograms; the average weight of females is 33.4 kilograms (Smith and Jungers 1997). Thus, nonreproductive females require about 20 grams of protein a day. This suggests that they can easily obtain half of their requirements through insect prey alone. Males and pregnant or lactating females have higher protein demands, but insects could make a significant contribution. Estimates of the body mass of australopithecines are very similar to those of chimpanzees. *P. robustus*, for instance, likely ranged from 32 to 67 kilograms across the sexes (Kappelman 1996).

This model reconstructs an average diet for these hominins based on the nutritional requirements of similarly sized extant apes. However, *P. robustus* would have encountered significant seasonal variability during the early Pleistocene. Their strong masticatory morphology suggests that they were capable of eating hard and tough foods. These foods were likely a fallback when nutrient-rich foods such as fruit were less available. Stable isotope analysis using laser ablation techniques takes small samples from the same tooth at different stages of enamel growth to assess variation in diet over the course of a single lifetime. Using this technology, Sponheimer and colleagues (2006) found that the diet of *P. robustus* varied both seasonally and annually, ranging from a predominantly C_3 diet to one dominated by C_4 resources. Even chimpanzees that live in savanna environments do not experience this extreme of seasonal variation in their diets. Baboons, however, may provide a better model. When fruit is low, they dig up the underground storage organs, or corms, of savanna grasses and sedges. These resources could have contributed to the C_4 signature seen in the hominin fossils (Codron et al. 2006). In every season, the majority of the plant-based part of the diet, whether it was fruit or corms, would have been low in protein. This evidence suggests that supplementing with protein-rich insects would have been beneficial to these hominins. *P. robustus* could have used bone tools to dig up corms in seasons when preferred resources were unavailable but could have used those tools all through the year to supplement their diets with protein-rich termites.

Female Australopithecines

Like the great apes, insect consumption by our early hominin ancestors probably focused on social insects. These edible insects, which include ants and stingless bees in addition to termites, provide an appealing option as a food source because they occur in colonies that number in the millions and live in nests that are often easy to find. Optimizing the harvesting process beyond the techniques chimpanzees use could have helped support the 20 percent increase in brain size seen in later australopithecines. The use of bone tools would have increased foraging success, including reaching fat-rich nymphs, larvae, and eggs, which not only increase caloric intake but also provide the essential fatty acids that are necessary for brain development and function.

Female australopithecines may have taken particular advantage of edible insects to the benefit of their dependent and semi-dependent young (Backwell and d'Errico 2008). Tanner and Zihlman (1976) have asserted that the role of the early hominin adult female was critical to the evolution of the human condition. Maternal investment at this time increased beyond that of chimpanzees, including expending the energy needed to carry infants while bipedal and caring for young during a longer period of dependence. The technological advances mothers made and their increased foraging efficiency were important for acquiring the quantities of food needed to support the nutritional demands of their families. These technological advances likely included not only the stone and bone tools found in the archaeological record but also other tools such as sticks for digging subterranean resources and for reaching arboreal foods. Foraging efficiency could have been increased further if mothers socialized their maturing offspring to contribute to the well-being of their siblings. Chimpanzees and other nonhuman primates contribute some to sibling care, usually by carrying infants, and the frequency of these contributions likely increased at this point in hominin lineage. Additionally, siblings could have contributed by provisioning food. A shift to communal foraging efforts is the foundation of the evolution of home bases and the nuclear family. This pattern is especially important for later hominins of the genus *Homo.*

* * *

Because we have archaeological evidence that *P. robustus* used bone tools for digging, it is likely that they benefited from consuming social insects. The insect portion of their diets can be reconstructed using models from great ape behavior with the understanding that these hominins likely needed higher-quality resources to support the increased brain sizes that developed later in the lineage. It is possible that insects were critically important at this point in human evolution, but they are not enough to explain how our brains kept increasing to the sizes we see today. In order for humans to achieve the magnificent powerhouse of a brain that we now have, we needed to evolve increasingly complex cultural systems. It is with the genus *Homo* that these behaviors became more critical to survival, and as a result, their culture, including their insect-eating practices, began to resemble the foraging patterns humans use today.

7

Edible Insects and the Genus *Homo*

Insects are used as food most often in tropical and subtropical environments. It is almost certain that australopithecines that lived entirely on the African continent incorporated insects into their diets. However, the evidence is not as clear for hominins of the genus *Homo* that expanded their range beyond the continent of Africa. It was in trying to conceptualize the insect portion of the diet of hominins in northern latitudes such as European Neanderthals that I began to search for correlations of insect eating with different habitats. In present-day populations, the likelihood of insects in the diet decreases with increasing distance from the equator (Lesnik 2017). The environment is a major factor that affects this food choice. This pattern can contribute to models of the likelihood of insect consumption by earlier members of the genus *Homo* based on their geographic location. In this chapter, I explore the shift to human-like use of insect food resources over the evolution of the genus *Homo*.

Evolutionary Origin of the Genus *Homo*

We belong to the species *Homo sapiens*: the sapient, or wise. We are the only surviving hominin species after a legacy of evolutionary diversity, including numerous species of australopithecines and multiple species of the genus *Homo*. The earliest fossils attributed to the genus *Homo*, which appeared in the fossil record over two and a half million years ago, belong to the species *Homo habilis*. The bodies of these hominins were small, similar in size to contemporaneous australopithecines, but they had a slightly larger, more rounded brain case and smaller dentition that bore very little resemblance to the teeth of more ape-like predecessors. The

Figure 7.1. Comparison of casts of hominin crania. *From left to right: Ardipithecus ramidus, Australopithecus africanus, Paranthropus boisei, Homo erectus* (Dmanisi), *Homo erectus* (Zhoukoudian), *Neanderthal,* present-day *Homo sapiens.* Photo by J. Lesnik.

name *Homo habilis* suggests these hominins were skilled or handy; they were given this name because they were found with the oldest stone tools that had been discovered at the time (Leakey, Tobias, and Napier 1964). We now know that australopithecines were capable tool makers and users, but the tools and associated butchering at the site of Olduvai, from which the Oldowan core-and-flake tool technology gets its name, are attributed to *Homo habilis* (Leakey, Tobias, and Napier 1964). Butchering sites, which became more common over time, reflect two important behavioral shifts within the genus *Homo*: an increased emphasis on meat in the diet and the establishment of home bases where meat and other resources were brought to process and to share.

Fossils designated to the species *Homo erectus* emerged about 2 million years ago (figure 7.1). These fossils are the first to truly reflect the human condition: their brains and bodies were larger than those of their predecessors, and thus a behavioral shift must have accompanied these morphological changes in order to support their energetic demands (Leonard and Robinson 1994; McHenry and Coffing 2000). I suggest that it is during this period that insect foraging began to look more like the foraging of present-day hunter-gatherers than that of chimpanzees or other nonhuman apes living today.

Homo erectus, Energetics, and the Division of Labor

Homo erectus was a long-surviving species. The oldest fossils of this species date to 1.9 million years ago and their presence in the fossil record continued for more than a million years after that. *H. erectus* was the first hominin species to leave the continent of Africa; fossils have been found as far to the east as Indonesia and as far north as present-day Georgia in the Caucasus and Beijing, China (Bar-Yosef and Belfer-Cohen 2001; Shen et al. 2009). Because of this great expanse of both time and space, there is a large amount of variability in the fossils designated to the taxon. Height estimates for individual *H. erectus* skeletons range from 4'9" to 6'1"; estimates of body size suggest that many of the individuals were well over five feet tall. Brain size also varies across the species; the earliest specimens had brains in the range of 600–700 cubic centimeters, while the brains of later species exceeded 1,000 cubic centimeters, twice the size of the brains of the earliest australopithecines. The brains of the later species fit within the range of brain sizes in humans today.

With the evolution of more modern body proportions, *Homo erectus* must have made additional dietary and behavioral adjustments to accommodate the increased need for calories. Estimates of the total energy expenditure of *H. erectus* suggest it was about 40–45 percent greater than that of australopithecines (Leonard and Robertson 1997). Higher-quality foods such as meat and a wider breadth of plant foods would have increased the variety of the diet and helped offset these increased energetic costs. Stable carbon isotope data from *H. erectus* fossils from East Africa does not differ significantly from that of australopithecines from southern Africa; this suggests that about 35 percent of their diets came from C_4 resources (Cerling et al. 2013). However, researchers theorize that meat contributed much more to the diet of *H. erectus* than it did to the diet of australopithecines. Microscopic wear patterns preserved on dental enamel is evidence of the changing diet of *H. erectus*. These patterns show that *H. erectus* had a generalized diet; they did not specialize in extremely tough or hard foods. Compared to what researchers have deduced from the microwear on the dentition of earlier hominins, the *H. erectus* diet appears to have been more variable across individuals, reflecting a diet that included a greater range of foods of varied hardness and toughness (Pontzer et al. 2011; Ungar and Sponheimer 2011; Ungar et al. 2012). The variation in dental wear patterns among individuals may reflect strong seasonality in the *H. erectus* diet. This is because dental microwear reflects what the individual was eating in the days or weeks prior to death. Thus, the microwear patterns in the dentition of individuals tell us what they were eating during the season they died (Grine 1986).

Although the earliest *H. erectus* specimens were found in association with tools similar to those the australopithecines made, by 1.5 million years ago a new, more complex technology had emerged that is known as the Acheulean (Klein 2009). This industry includes tools such as flakes, cleavers, and scrapers, but it is best known for its hand axes—large stone cobbles worked symmetrically across the midplane, or "bifacially flaked." To make these tools, *H. erectus* must have had a predetermined form in mind and used an extensive sequence of knapping techniques to systematically remove flakes to achieve a goal. The foresight and technical skill with which these tools were created are orders of magnitude greater than what was needed to fashion the earlier Oldowan technology. The utility of these hand axes must have been great because they persist relatively unchanged in the archaeological record for over a million years. Archaeologists theorize that *H. erectus* likely used these tools for a variety of purposes, from butchering animals to chopping down trees (Dominguez-Rodrigo et al. 2001; Solodenko et al. 2015).

Chemical analyses of organic residues preserved on the edges of some of these tools confirm that they were used to process both plant and animal foods. Residues of animal fat were found preserved on Acheulean tools at Revadim Quarry in Israel (Solodenko et al. 2015). Abundant elephant bones were also found at the site, including a rib with cut marks. *H. erectus* clearly used the hand axe to process animals in some way, and they likely used it to butcher elephants. Three hand axes were analyzed for the plant phytolith content preserved on their surfaces at the site of Peninj in Tanzania (Dominguez-Rodrigo et al. 2001). Phytoliths are microscopic silica structures that are stored in plants. The morphology of phytoliths indicates the plant type from which they came and can even be assessed to the species level in some cases. Dominguez-Rodrigo and colleagues determined that the phytolith content of the soil matrix surrounding each of the three hand axes was significantly different from the phytolith content found on the artifacts. This confirmed that the residue on the tools was a by-product of use, not contamination. The phytoliths in the residue strongly resemble those of woody plants; they were best matched with *Acacia* trees. This finding, along with the large amounts of damage on the edges of the tools, suggests the hand axes were used for woodworking. The authors speculate that the hominins may have fashioned spears for hunting, but they just as easily could have been stripping wood to make baskets.

A carrying vessel such as a basket or bag made from animal hide would have increased the foraging efficiency of *H. erectus* tremendously. Making such a vessel would easily have been within their mental capacity. Linton (1971) has suggested that an invention of this type would have been one of the earliest and most important in human evolution. Females were likely the inventors of these early hominin tools, since the increased returns from their efforts supported their reproductive success both indirectly by reducing their energy expenditure and directly by increasing their ability to share collected food items with their offspring (Tanner and Zihlman 1976; Zihlman and Tanner 1978; McGrew 1981). Zihlman (1978) argues that gathering and hunting must be considered as interrelated. The home bases that appear in the archaeological record alongside the fossils of early *Homo* imply that a higher level of community organization evolved with greater cooperation than was seen for earlier hominins. The foods *H. erectus* females gathered and brought back to home base could be shared not only with their offspring but also with adult males, initiating communal exchange and the assurance that all individuals would eat. These social ties would have enabled adult males to venture farther away from home base to search for resources that involved greater risk, such as large game or raw materials, because even if they failed, there would still be food for them at camp when they returned.

Beyond the shared resources that may have been typical of *H. erectus* society, it is also possible that the subsistence activities females conducted were optimized for their nutritional needs and that they would have collected insects and other small animal foods in addition to plant-based foods. In modern foraging societies, women tend to collect and consume more insects than men. In latitudes closer to the equator, social insects are consumed year-round and other seasonally abundant insects are valued when they become available. It is highly probable that *H. erectus* females living in tropical regions devoted time every day to finding and collecting insects. Because of their increased brain size and their advanced tool technology, these hominins had the ability to exploit seasonal insect resources at a higher level than australopithecines did. These resources would have been particularly useful to *H. erectus* living at latitudes farther away from the equator, where large colonies of social insects are less common. Modern foragers take particular advantage of beetle larvae and caterpillars when they are seasonally available. Caterpillars often appear in massive numbers at once. Earlier hominins may have opportunistically enjoyed eating these insects, just as a lot of nonhuman primates do, but *H. erectus* would have benefited from collecting large numbers of them in containers so they could share them with other members of their group. Beetle larvae are another insect resource that modern foragers value highly. Heavy tools are needed to extract these insects from trees and logs. This technology was unavailable to earlier hominins, but it is possible that *H. erectus* had access to it through the invention of the Acheulean hand axe. The opportunistic collection of other solitary insects such as grasshoppers, crickets, and cicadas would also have helped *H. erectus* obtain the nutrients necessary for reproduction such as folic acid, iron, protein, and fat. The diet of the species *Homo erectus* would have incorporated a large number of resources because they were dispersed over such large amounts of both time and space. The insect portion of their diets would have varied across populations, just as it does for modern hunter-gatherers.

In addition to making advances in tool use, another way to intensify resource acquisition from the environment is to cook foods, which makes more nutrients accessible and reduces toxins that might otherwise be harmful (Stahl 1984). Although evidence of fire exists in hominin sites as far back as 1.6 million years ago (Bellomo 1994), the earliest convincing evidence for hominin control of fire comes from the site of Gesher Benot Ya'agov, a 790,000-year-old Acheulean site in Israel (Goren-Inbar et al. 2004). The site is commonly attributed to *Homo erectus* since they are most associated with the Acheulean industry; however, the authors note that there is no evidence of a particular hominin species and that the occupants could have been archaic *Homo sapiens*. Thus, we still have no direct evidence that *Homo erectus* was capable of cooking food. Instead, credit for that behavior is reserved to later members of the genus *Homo*, such as Neanderthals, where there is abundant evidence of fire pits and cooking.

A Probable "No" for Neanderthals?

The fossil record shows that *Homo erectus* expanded their range across Asia but there is no indication that they ever occupied regions under glacial cover (Antón 2003). Not until the arrival of the Neanderthals, an informally defined group of fossils known from sites across Europe, the near and middle East, and western Asia and dating from 150,000 to

27,000 years ago, do we begin to see how hominins coped in the cold of more northern latitudes (Stanford, Allen, and Antón 2011). These hominins were sophisticated tool makers who used caves and fire, allowing populations to survive in the harshest conditions. The diet of the Neanderthals is most commonly modeled as meat-based. It can be difficult for humans to obtain all the nutrients they need from plant-based diets in environments with harsh winters. Our digestive systems are suited for high-quality foods that yield high nutritional content per unit of weight. Many other mammals regularly consume low-quality foods such as grass, leaves, and bark that are dense in cellulose that is impossible (or almost impossible) for humans to digest. Also, the variability in the availability of food resources is extreme in temperate zones. When winter arrives, fruit, seeds, and other high-quality plant foods are essentially unavailable. What remains is the animals that are able to feed on the low-quality foods that are unavailable to us. Hunting these animals is an ingenious way of extracting high-quality food resources from an environment that is otherwise inhospitable for humans. In addition to flesh, Neanderthals may have eaten the stomach contents of the animals they hunted, which would have provided carbohydrates through partially digested plant matter made available by fermentation in the stomach of a deer or other animal (Buck and Stringer 2014).

Stable isotope analyses of nitrogen and carbon preserved in Neanderthal fossils give an estimate of how much meat was in their diets. Because nitrogen in the body is provided by the protein portion of food consumed and carbon reflects the totality of the diet, the ratio of nitrogen to carbon can indicate the relative amount of protein an individual consumed. From this data, researchers can draw conclusions about the amount of meat in an individual's diet. Multiple studies of stable isotopes for Neanderthals show large amounts of nitrogen (Richards et al. 2000; Bocherens et al. 2005; Balter and Simon 2006; Richards and Schmitz 2008; Hublin et al. 2009; Richards and Trinkaus 2009). One issue with these studies is that compared to data for carnivores, the data for Neanderthals suggest that they had more protein-enriched diets than even hyenas and wolves. Of course, Neanderthals did not eat more meat than these carnivores; they were omnivores like all members of the genus *Homo* (Henry, Brooks, and Piperno 2011). The signals from the isotopes are the effects of different sources of protein and, in some instances, different locations, all of which produce variations in the amount of nitrogen available in the environment.

The results support the hypothesis that Neanderthals had a meat-based diet, but much more work needs to be done to understand how different plants and animals, both terrestrial and aquatic, selectively store different nitrogen and carbon isotopes. Without this information, we will not know the actual amount of meat Neanderthals consumed (Schoeninger 2014).

Eating insects is another way of converting unavailable resources into high-quality food. Like the game Neanderthals hunted, many insects feed on low-quality plant foods. Some insects would have been seasonally abundant in regions occupied by Neanderthals, such as the migratory locust (*Locusta migratoria*) in Europe and the desert locust (*Schistocerca gregaria*) in the Middle East. These insects have a deep history of being used as human food in these regions (DeFoliart 2002), and Neanderthals similarly could have enjoyed a reprieve from hunting when they were available.

During the harsher seasons, insect foods would have been difficult to find in amounts abundant enough to make seeking them out worthwhile. However, reindeer, a common prey for Neanderthals (Dusseldorp 2011), are hosts for the parasitic warble fly (*Oedemagena* or *Hypoderma*) (Nilssen and Gjershaug 1988). Larvae hatch from eggs laid in the hair of reindeer during the summer and then penetrate the skin, where they reside for months until emerging in adult form in the spring. Reindeer infected with the parasites are made weak and are easy targets for hunters. Larvae found while butchering reindeer during the winter are a valued food source for the Tlicho people of the Northwest Territories of Canada (Felt 1918) and could have been utilized by Neanderthals in the same way. Paleolithic art created by early *Homo sapiens* who occupied Europe after the Neanderthals depicts small wormlike creatures that Guthrie (2005) suggests represent these larvae.

Temperate areas have extremely low species richness compared to tropical ecosystems. This phenomenon is known as the latitudinal gradient in biodiversity (Boyero 2012). The number of species that inhabit a given place is highest in areas that have persisted largely unchanged for a long time and are characterized by predictable environmental conditions that do not deviate substantially from the long-term norm. When the Neanderthals existed during the late Pleistocene, up to 30 percent of the earth's surface was glaciated, much of it in the northern hemisphere. In what is now Europe, ice covered present-day Scandinavia, extended south and east across present-day Germany, and southwest to what we now know as the British Isles. Records of plant and animal remains (including insects) during the height of glaciation clearly demonstrate that most of the organisms presently distributed across Europe were in refuge in the south, especially on the peninsulas of Iberia, Italy, and the Balkans (Hewitt 1999). An analysis of insect diversity in northern Europe in our modern climate suggests that the number of insect species is highest in the most southern regions and that on the Swedish island of Öland, species found in habitats with poor, low-growing vegetation have been there since Late Glacial times (Väisänen and Heliövaara 1994). These studies taken together suggest that Neanderthals at the southern end of their range in the Middle East would have had more diversity in the foods available to them, including a larger variety of insects. Closer to the glacial cover, this diversity would have been much lower because fluctuations of glacial advance and retreat create unpredictable conditions that decrease the likelihood of species colonization.

Direct evidence of insect consumption is scarce across the entirety of the fossil record, but with the Neanderthals and contemporaneous modern humans, we begin to see evidence of the consumption of other invertebrates. Shells of marine mollusks found at the 150,000-year-old Neanderthal site Bajondillo Cave in southern Spain exhibit fracturing and burn marks consistent with the consumption of shellfish by Neanderthals (Cortés-Sánchez et al. 2011). At roughly the same time, modern humans at Pinnacle Point Cave on the coast of South Africa were also taking advantage of shellfish (Marean et al. 2007). Routine use of shellfish becomes common in South Africa after this time, as demonstrated by the preservation of shell middens, but for Neanderthals, this sort of evidence has yet to be found (Klein and Bird 2016).

While terrestrial mollusks such as snails are routinely found at Neanderthal sites, their presence is generally thought to reflect their natural use of caves. The earliest accepted evidence of snail consumption in Europe, which comes from the site of Cova de la Barriada on the eastern Iberian Peninsula, dates to 31,300–26,900 years ago (de Pablo et al. 2014). Here, archaeologists found over 1,000 specimens of *Iberus alonensis* in association with hearths, stone tools, and the remains of other fauna. This finding is truly spectacular and provides definitive evidence that Neanderthals consumed snails. However, it is likely that Neanderthals typically consumed snails at much smaller scales. Archaeologists tend to ignore snail shells at archaeological sites because they are thought to be naturally occurring, but these shells may be evidence of snail consumption. It may be worth revisiting these shells with this hypothesis in mind.

Modern Human Origins and Dispersal

The exact origin of *Homo sapiens*, or "modern humans," as they are often called, is difficult to determine. First, is the definition based on anatomical or behavioral modernity? Second, how many modern features must be present for a fossil or a fossil site to be classified as modern human? For any definition that is created, an exception can always be found. This lack of precise definition is only one problem. Another issue is a major division among paleoanthropologists about which theoretical framework to use to interpret data relevant to this question. Some scholars posit that there is a single point of origin for anatomical modernity that occurred in Africa about 200,000 years ago (Tattersall 2009). This hypothesis suggests that modern humans dispersed out of Africa, ultimately outcompeting and replacing the other hominin species that were already present throughout the Old World. The result was that *Homo sapiens* was the only hominin that survived. An alternative view suggests that there is no single origin of modernity and that instead modern humans are the result of continuous genetic admixture among ancient populations across the Old World (Wolpoff et al. 2001). In this model, the evolution of a beneficial modern trait in one region would quickly spread across the Old World through the mechanisms of gene flow and natural selection. From this point of view, we are not the last hominin left standing but instead carry with us genetic information about our ancestors that reaches back to the original dispersal of *H. erectus*.

These two models generate different interpretations for almost all behaviors for this time, including the antiquity of human insect eating. Today, throughout the Old World, the most common locations where insects are consumed are in tropical and semitropical areas, which happens to be the range *Homo erectus* occupied. Paleoanthropologists who subscribe to the admixture model of modern human origins would find it probable that insect-eating behaviors outside Africa persisted in these regions from the time of *H. erectus* and that knowledge about edible insects in places like Indonesia is part of the collective memory of that region for as many as a million years. Proponents of the admixture model look to fossil evidence of early modern humans at the peripheries of the human range. If the replacement model is correct, all of these fossils, whether they were in Europe or Australia, should be more similar to older fossils in Africa than to fossils from the more ancient populations, such as Neanderthals or Indonesian *Homo erectus*, that they were thought to replace. However, studies by Hawks and colleagues (2000) and Wolpoff and colleagues (2001) demonstrate that regional traits persist over time and thus refute a model of complete replacement. Recent sequencing of the genome of present-day humans demonstrates that we have genes that evolved in ancient populations outside of Africa such as the Neanderthals. This data confirms that some amount of genetic admixture occurred between the earliest *Homo sapiens* and previous hominin populations (Prüfer et al. 2014).

Behavioral modernity, which arises much later than anatomical modernity, is often defined by the presence of delicate tools; an increased variety in hunted foods, including aquatic resources; and symbolic behavior including elaborate art and decoration (Klein 1995; Richards et al. 2001). Some of the earliest evidence of these behaviors have preserved in coastal sites in South Africa. One particularly rich site is Blombos Cave on the Southern Cape, which dates to 70,000–100,000 years ago. Material culture found at the site includes ochre, engraved bone, shell beads, and points made of both stone and bone that were likely fashioned into spears (Henshilwood et al. 2001; d'Errico et al. 2005). At this site, the fossil remains of terrestrial game and a large range of aquatic resources, including large fish, shellfish, seals, and dolphins, have preserved. Although edible insects would be nutritionally redundant in this large range of animal foods, they may have offered a resource at times when preferred resources were less available or harder to obtain. Contemporaneous inland populations without easy access to aquatic resources likely relied on insects more heavily.

Over time, modern humans colonized the far reaches of the world. Like Neanderthals, Arctic populations could not have survived the extreme cold without depending on hunting or fishing. In these colder areas, insect diversity is rather low, so these populations probably did not consume insects in quantities greater than what we see today in that region of the globe (Jongema 2017).

Homo erectus traveled to what are now the islands of Indonesia when they were connected to the continental mainland during the Pleistocene (Antón 2003). During the glacial advance of the Ice Age, sea levels dropped and the continental shelf of Asia created a continuous landmass out of the islands of southeast Asia. However, a deep-water channel still remained between this continent and the landmass of Oceania. In order for hominins to reach Australia, they had to traverse these waters. The fossil record suggests that this was not achieved before the arrival of modern humans (Thorne et al. 1999). Thus, insect consumption in Australia is a recent occurrence in human evolution.

The peopling of the Americas was the last major expansion of the modern human range. Although researchers debate whether the first people arrived by land or by sea, recent dates from the site of Monte Verde in Chile suggest that humans occupied coastal South America as far back as 18,000 years ago (Dillehay et al. 2015). The two models for the arrival of humans suggest different theories about the origins of insect consumption in the Americas. If people arrived solely by way of the Beringia land bridge from Siberia, they would have been accustomed to a tundra environment and would not have included significant amounts of insects, if any, in their diets. From this perspective, insect consumption would be a recent acquisition in the New World just like it is in Australia. However, if people traveled by boat from Polynesia, they would almost undoubtedly have recognized insects as edible upon their arrival. Regardless of who got there first, the pattern of insect consumption across the Americas today has been shaped by the behaviors brought via these two pathways.

The general pattern of insect consumption across the Americas matches that of the Old World: the highest frequency of insect eating happens in the tropical areas, while eating insects is rare at higher latitudes, both north and south. In the archaeological record, some evidence of insect eating in the past has been preserved in coprolites, desiccated or mineralized feces that are preserved in shelters or in open sites in arid regions. These are found primarily in the New World. Insects have been recovered from human coprolites at sites in Mexico, Peru, and the United States. Today, people in Mexico are among the top consumers of insects worldwide, and there is likely a deep history for this practice. Coprolites from caves in the Tamaulipas region of northeast Mexico indicate that in the span of time from 8,700 to 450 years ago, people there consumed insects such as grasshoppers, caterpillars, ants, flies, beetles, bees, and wasps (Callen 1965).

The most abundant evidence for insect consumption in the New World comes from coprolites found in the United States. Sutton (1995) reviews this archaeological evidence of insect consumption. Many caves in the Great Basin area, including sites in present-day California, Oregon, Utah, and Nevada, have yielded insect remains. Other edible insect evidence comes from coprolites found in present-day Arizona, Colorado, Texas, Kentucky, and Arkansas. All of these sites are in semiarid steppe, mid-latitude desert, or subtropical climates. The arid environments of deserts and steppes are most conducive to the preservation of coprolites, but coprolites preserved in caves and rock shelters in the more humid subtropical sites in Kentucky and Arkansas demonstrate that insects were consumed in that climate as well. Some of the evidence for insect eating in these coprolites is in the form of the nonspecific presence of insect chitin. However, some important specimens related to the human consumption of insects exist. In the Glen Canyon area of southern Utah, coprolite samples that span a sixty-five-year occupation show an increase in insect remains over time (Fry 1978). At Dirty Shame Rockshelter in southwest Oregon, termites of *Reticulitermes* made up 78.3 percent of one of the coprolites (Hall 1977). At Bamert Cave in east-central California, crane flies made up 25 percent of a coprolite (Nissen 1973).

When we compare insect eating in the United States today to the evidence of this practice preserved at these archaeological sites, it is clear that a lot has changed in food culture since the arrival of the first peoples to the New World. The world has gone through two major food revolutions in the past 10,000 years. The first was the result of the adoption and spread of agriculture and the second was the result of industrialization. With each of these transitions, increasingly less emphasis was placed on collected foods. Skeletal evidence for early agricultural populations suggests that the adoption of cultivated foods corresponds with a decline in oral and general health (Larsen 1995). Dietary variety was very narrow, which resulted in nutritional deficiencies. One useful way to both protect crops and increase intake of key nutrients is to gather insect pests, such as grasshoppers, as food. The benefit of using this natural method of controlling agricultural pests is evident today in its independent use in populations across the continents (Cerritos Flores, Ponce-Reyes and Rojas-García 2014; Niassy et al. 2016). In one example, these insects have become a symbol of identity. In Oaxaca, Mexico, *chapulines*, or toasted grasshoppers, are used as symbol of Oaxacan identity (Thrussell 2016).

In the Zapotec communities of rural Oaxaca, the *chapulineras* are women who collect grasshoppers from agricultural plots and sell them in local marketplaces (Cohen, Sánchez, and Montiel-Ishino 2009). In recent years, there has been a shift in the structure of sales of *chapulines*; fewer sales are being made at the household level but more *chapulines* are being sold to bulk buyers who sell them at sporting events or other cultural affairs that attract tourists. This reduction in household consumption of insects is a result of industrialization. Younger generations of Zapotec have forgotten or never learned how to manage a farm, and thus do not value the edible insects that are a by-product of pest management (Wilk 2006). Store-bought packaged foods are increasingly common even in rural areas of Mexico, and traditional foods are gradually being phased out in favor of products available on the global market. This example from Oaxaca illustrates the modern use of insect foods. What once had a long-standing tradition, possibly reaching back to the first inhabitants of the region, is now being replaced with lower-quality foods that are produced on the industrial scale.

* * *

The genus *Homo*, going back to *Homo erectus* and continuing through the origins of modern humans, likely foraged for insects, most commonly in tropical regions. Like modern foragers, *Homo erectus* females would have benefited the most from securing this resource. Technological innovations, such as a carrying vessel, would have enabled these hominins to bring abundant quantities back to a home base to share with their families.

With the evolution of *Homo sapiens*, humans began to occupy regions far outside the tropics, such as Europe, and over the next 100,000 years, human populations radiated all across the world. Innovations in food procurement, such as fishing nets and domesticated crops, changed how people ate; they began relying less on foraged foods and more on skilled labor. This trend continued all the way through the industrial revolution, leading to where we are today. Although insect consumption today is still most common in the tropics, the presence or absence of this food in any culture can only be truly understood through their local histories. We need to continue discovering ways to reveal more about how edible insects were used in the past.

8

The Potential for Future Discovery

Testing Hypotheses of Edible Insects

The process of scientific discovery works in two ways: induction and deduction. In archaeology and paleoanthropology, a fair amount of induction happens as artifacts and fossils are discovered that need explanation. For edible insects, induction would look something like this: "There are bugs present in this coprolite: people were eating bugs." Deduction is a far more complex process of hypothesis testing. Hypothesis testing is the heart of the scientific method, which includes making observations, forming hypotheses, conducting experiments, collecting and analyzing data, and drawing conclusions. However, it can be difficult to make sense out of observations and create testable hypotheses. The purpose of this book is to try to make sense out of observations related to diet and insect eating in extant populations of people and nonhuman primates and in the archaeological record. In the previous chapters, I applied these observations to our hominin ancestors and created models of how insects were likely used in their diets. These models set up the following hypotheses, or predictions, that we may be able to test as more direct evidence becomes available:

1. Early hominin diets included significant amounts of insects that belong to social taxa such as termites and ants.
2. Female hominins ate more insects than their male counterparts.
3. In the genus *Homo*, insect consumption occurred more frequently in tropical regions than the regions distant from the equator.

4. More edible insect options became available as members of the genus *Homo* created technology useful for harvesting insects, thus increasing the diversity of this portion of their diet.
5. When sexual division of labor occurs in a society, it is the women who tend to focus on the collection of insects and consume more of them.

These predictions are not easy to test. The difficulty with paleoanthropology is that researchers work with scant evidence. However, there are two important ways this research can advance: first, by continuing to collect relevant data from extant populations to refine these models; and second, by identifying what to look for in the paleoanthropological record that can be used to test these ideas.

Standardization of Data from Extant Populations

In the preface of his book *Insects as Human Food*, Bodenheimer (1951) states:

> The factual material on this subject is so enormously scattered in journals and books on travel, ethnology, geography, medicine, zoology, etc., etc., that it is an almost impossible task to aim at gathering all available facts. (5)

Because each field of study has its own framework for reporting and interpreting how people incorporate insects into their diets, it is difficult to synthesize information to use for testing hypotheses regarding cross-cultural patterns of insect eating, such as differences between sex or age groups.

Data regarding insects as food is much more limited than it is for other resources because of a western bias against insect consumption. In anthropology, cross-cultural comparisons are important. Because of this, large databases such as the Standard Cross-Cultural Sample have been created, but none of these databases has a category dedicated to insects as food. If insects appear at all as food, they are categorized with other small-sized animal foods that are foraged, such as eggs, other invertebrates, and small vertebrates. When insects consumed as food are not coded as a separate variable, the data in these databases cannot be used to analyze insect consumption across studies. Data dedicated specifically to insects as food are needed in order to collect the information we need to generate and refine models of how they are used worldwide. Although additional studies that focus exclusively on insect consumption would be of great value, much could be gained if researchers doing broad food surveys would code insects as a separate variable.

Because the nutritional values of edible insects vary greatly across the diversity of species, reporting only the broad category of insects in the diet is not enough. The most useful studies break down the insect portion of the diet to the different species, listed by scientific taxon and by the life stage they are in when consumed. These methods tend to be used only when insects are the specific focus of the research, and even in these studies, reporting is not always consistent (Raubenheimer and Rothman 2013; Raubenheimer et al. 2014; Rothman et al. 2014).

Analysis of nutritional content is often also unstandardized. Sometimes this task is as simple as converting a value reported as the proportion of fresh weight to the proportion of dry matter or converting the unit of measure from milligram per gram to gram per one hundred grams. However, sometimes the methods used to generate the value for the amount of a nutrient differ and simple conversions cannot be used to make values comparable for a larger body of work.

We have an abundance of literature on insect foraging by chimpanzees, but these data often stem from an interest in tool use, not from an interest in insects as food. The data on edible insects are much more limited for species that do not use tools. When researchers observe small-bodied primates that forage for solitary insects, it can be difficult for them to see insect-eating events and report observations with confidence. However, the research led by Amanda Melin (Melin et al. 2010; Melin et al. 2014) on insect foraging by capuchin monkeys is exemplary of how this data can be collected when it is the primary goal of the research. Melin and colleagues note that "our methods for recording invertebrate foraging differed from those of recording fruit foraging because, in contrast to fruit tree visitations, invertebrate faunivory is relatively cryptic and requires intense observation of one individual at a time to witness and identify feeding records" (2014, 79). With these methods, they were able to address age and sex differences in preferred prey, time allocation for insect collection, foraging efficiency, and seasonal variations. More projects like this with other primates across the order will help us understand how different primates incorporate insects into their diets.

It is not only nonhuman primates but also present-day human populations that use insects as food differently by age and sex categories, yet we do not know much about these patterns. We need more data in these categories if we are to understand the variables that may affect these patterns. Insects are likely consumed across all age and sex categories within a species, but not necessarily in the same amounts. Because insects are a nutrient-dense food, slight variations in the amounts consumed affect how they contribute to the dietary needs of individuals. Refining reports of edible insects to include the different age and sex classes of consumers will be critical to understanding the nutritional niche insect foragers fill. Research should aim to identify patterns of insect consumption, such as who catches them, how are they prepared, and who eats them.

Analytical Methods of Diet Reconstruction

In 2001, McGrew authored a chapter in *Meat Eating and Human Evolution* titled "The Other Faunivory: Primate Insectivory and Early Human Diet." In 2012, he revisited this topic by co-organizing a session with O'Malley for the annual meeting of the American Association of Physical Anthropologists that resulted in a special issue of the *Journal of Human Evolution* (O'Malley and McGrew 2014). The aim of McGrew's "other faunivory" is "to explore the likely importance of the non-vertebrate potion of early human diet; to what extent (if at all) did the earliest *Homo spp.* eat invertebrates" (2001, 161) and to offer hypotheses for archaeological testing. McGrew's opinions echoed those of archaeologist Mark Sutton in 1995:

> To begin to deal with insect fauna from archaeological contexts, archaeologists must first develop an appreciation for the role of insects in past cultures, as indicators of environment, and in site formation and taphonomy. Field and laboratory methods must be sensitive to recovery and analysis. Archaeologists must expect to find insects in sites and must approach insect fauna in the same way as they do other fauna. (288)

Here I provide my perspectives on the future of edible insect research in paleoanthropology. While it is crucial to consider insects as a possible food source in archaeological contexts, some of our methods do not give us the capacity to detect insect consumption from the available evidence. Another issue is that we lack the data from extant contexts with which to

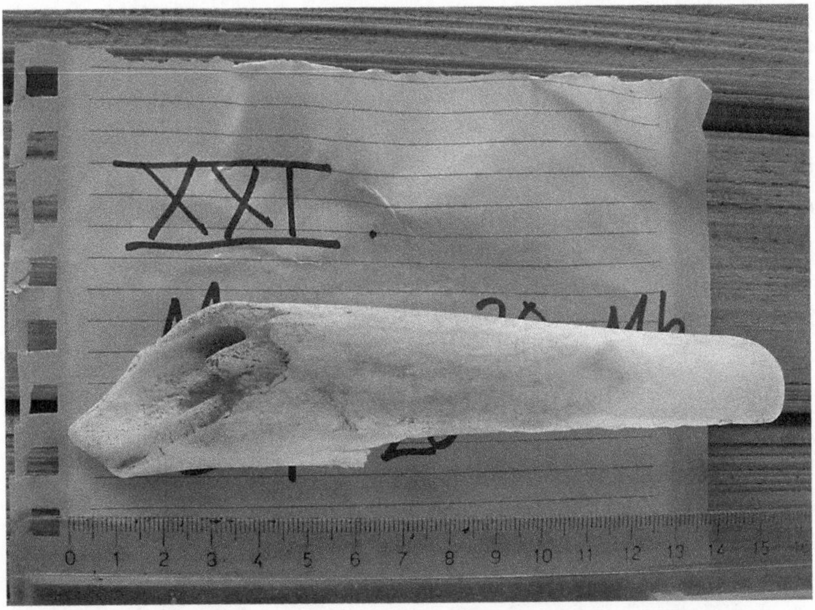

Figure 8.1. Experimental bone tool made from a freshly butchered cow (*Bos* spp.) tibia used to dig into a termite mound for a total of 40 minutes. Wear and polish on the working end are easily visible after this amount of use and are comparable with the wear seen on the artifacts from South African *Paranthropus robustus* sites such as Swartkrans. Photo by J. Lesnik.

compare archaeological finds. I believe there are many interesting, testable research questions that will help create an important body of knowledge that is necessary for understanding the role of insects in the hominin diet. Below I discuss several methodologies and their potential contributions to research on insects as food.

Tools and Use Wear in the Archaeological Record

The bone tools like those from the Swartkrans site provide the evidence necessary for paleoanthropologists to have serious conversations about hominin insect eating (figure 8.1). In 2001, McGrew predicted that analysis of microwear and polish on tools had the potential to identify how they were used to process (chop, slice, grind) invertebrate prey, although he focused on stone tools, which are more common in the archaeological record. Although studies of high-magnification analysis of microwear on lithic tools have been criticized (Newcomer, Grace, and Unger-Hamilton 1986), concerns regarding observer subjectivity are being addressed with the improvement of analytical methods (Stemp, Watson, and Evans 2015). Of course, good research design is critical to the success of microwear studies. As Keeley has noted: "There are undoubtedly some disappointments in store for microwear analysts, since there are probably some questions about tool use that microwear analysis will never be able to answer, but these true limits will not be reached through the use of haphazard techniques and methodologies" (1974, 334). McGrew suggests that the chitin and calcium carbonate present in insect bodies may leave a distinct wear or polish on tools but admits that it remains unknown if these patterns would be discernible from other polishes (McGrew 2001).

As I began my study of use wear on bone tools (Lesnik 2011), I was quite confident that 1) bone is more malleable than stone, so is more likely to hold distinct patterns; 2) the architecture of *Macrotermes* and *Trinervitermes* mounds was so different that they would produce different patterns of microwear on tools used to penetrate them; and 3) advancements in use-wear analysis that make it free of observer error and scale-sensitive would identify these differences. However, after experimentation and data analysis, I found that my predictions about these different mounds could not be supported. Yet the results still provide insights, including insights into the limitations of use-wear analysis. For one thing, there is a real possibility of a false positive: what appears to be a match in the use wear on an experimental control and the archaeological artifact may not actually be a true match. It remains important to heed the critiques of microwear analysis that suggest polishes from contact with different materials can overlap (Newcomer, Grace, and Unger-Hamilton 1986). Researchers studying the South African bone tools acknowledge that the wear patterns preserved on these artifacts are variable and exceed the range of variation present in the patterns preserved on the experimental tools used to dig into termite mounds (Backwell and d'Errico 2008; Lesnik 2011). Possibly more important is that use-wear studies are useful for *rejecting* hypotheses. For instance, Backwell and d'Errico (2008) found that the wear on bone tools from Drimolen was significantly different from that on tools used to dig tubers but that this difference did not always reach the same level of significance for the Swartkrans tools. This finding identified an important difference, either behavioral or environmental, between the sites. We currently do not know what sort of wear pattern, if any, digging into a termite mound with a stone tool would yield, even though this has been demonstrated as an effective method of foraging (Lesnik and Thackeray 2007). Nor do we know what processing insects, such as grinding beetles or ants into a paste (Nonaka 1996), would do to a stone tool. Conducting such experiments would make it possible to look for more for evidence of insect consumption in the archaeological record.

Residues of Paleo-Foodstuffs

An abundant microfossil record exists that includes the residue preserved on artifacts. Paleoethnobotanists, or researchers interested in the plant remains that are present in archaeological contexts, are experts in analyzing the microscopic morphology of various structures of plants, such as starch grains and phytoliths, both of which preserve well long after the plant has died and decayed. Phytoliths are especially diagnostic, but it can be labor-intensive to identify phytoliths to the species level, as it is impossible to know what every plant species today looks like or what all plants in the past looked like. Most researchers are able to easily identify starch grains and phytoliths to the family level, and when the research question necessitates it—for example, if a researcher wants to identify early domesticates—they can be identified at the species level (Piperno 2006).

Similar analyses can be done with residue from animal products. At a general level, fat, protein, bone, hair, fish scales, feathers, and so forth can all be identified. This information is enough to tell whether a tool was used to process animal foods, for instance. Unfortunately, animals do not have a phytolith analog, something that preserves well and can be identified through microscopic analysis. However, biomolecular components of animal residue can sometimes be used to identify what are called chemical fingerprints (Evershed 2008). Analyzing these chemicals is a complex process and the application of the results is still limited because we lack a rigid framework for testing hypotheses. However, I believe these chemical fingerprints have great potential for identifying insect foods. Is there a chemical fingerprint for chitin that could be used in residue analyses? Do insect fats look the same as all animal fats? These are worthwhile questions to investigate.

Many taxa of edible insects have the capacity to yield unique biochemical fingerprints. For instance, many termites produce chemical toxicants, and even termites with primarily mechanical defenses, such as *Macrotermes*, produce small amounts of these defensive substances. Entomologists have identified species-specific chemical signatures for these substances (Kaib, Brandl, and Bagine 1991), which may offer a way to distinguish residue left behind on tools from processing termites from residue left by other animals and even by other insects. By understanding fingerprints of modern termites to the species level, we could make more accurate assessments of their nutritional contributions if such residue was found in archaeological contexts.[1]

One place to look for such residue would be on the ends of the bone tools found at South African sites. Although these tools were not used to process the termites directly, it is possible that the termite mounds the tools were used to excavate also preserve the species-specific chemical signature in their sediments. This signature may have been transferred to the bone tools as well. Unfortunately, these artifacts have been cleaned numerous times for the purpose of wear pattern analyses, so any ancient residue has long since been destroyed. As residue analysis is becoming more common, methods are increasingly being used in the field to protect potential residue from contamination. Simply wearing gloves and wrapping an unwashed artifact in aluminum foil before placing it into its own paper bag is generally enough.[2] If future research yields more bone tools in South Africa, collecting and storing the artifacts in this way would allow for future residue analysis.

Analysis of dental calculus is an area of research that holds great potential for identifying the residue of insect foods. Calculus, or hardened dental plaque, preserved on fossil teeth can reveal microscopic evidence of the foods consumed. This type of research has revealed that Neanderthals used fire to cook plant foods; starch granules that were clearly modified by heat were found in the calculus of Neanderthals from Shanidar Cave in Iraq (Henry, Brooks, and Piperno 2011). There is great potential for future research to identify insect elements in hominin dental calculus. Unfortunately, until recently, calculus was often scraped off when fossils were cleaned. It is hard to think about all of the information about diet that was lost because of these practices, but current standards are improving the preservation of dental calculus now that its analytical value is known.

Dental Microwear

Dental microwear studies, like studies of use wear on artifacts, have come a long way in recent years due to advances in analytical methods. Because dental enamel preserves microscopic pits and scratches, fossil dentition can be compared to modern samples with known diets. The issue with dental microwear, however, is that only general categories related to the mechanical properties of food can be identified. In primates, frugivores and folivores present very different microwear patterns, but in omnivorous taxa, such as fossil hominins, we cannot identify which exact foods were being consumed. For omnivores, the data can offer an indication of the amount of variety in the foods consumed based on the complexity of the microwear features. If they are all roughly the same, then most of the diet came from a single food type, but if there are lots of different-looking features, that suggests variety in the diet. For instance, based on analysis of dental wear, researchers believe that *Homo erectus* had a more varied diet than earlier australopithecines. However, neither species' dental microwear can yet say anything about the presence of insects in the diet.

Microwear analysis of the dentition of myrmecophagous mammals—animals such as aardwolves that specialize in eating ants and termites—suggests that their faunivory leaves a relatively distinct pattern characterized by a high frequency of pit features (Strait 2014). Thus, it seems highly probable that insect specialists can be identified based on their dental microwear. The problem with applying this finding to omnivorous hominins is that the pattern associated with insect consumption is easily masked by the scratches and scars that accumulate from eating a range of other foods.

It is likely that the limitations of dental microwear analysis will prohibit it from identifying the supplemental use of insects as food, such as the limited scale of insect consumption among chimpanzees and what was likely for our hominin ancestors. However, research testing for differences in dental microwear associated with varying degrees of insect consumption has yet to be done. One way to address this question is to study the dental microwear from different chimpanzees with known amounts of insect consumption. My hypothesis is that the dental microwear on the teeth of a very committed termite forager looks different from that of an individual who ignores termites. Skeletons of known wild chimpanzees are often collected for additional study at long-term research sites, and it would be a straightforward process to look for a correlation between an individual's frequency of termite foraging and their dental microwear patterns. A study such as this would also be useful for addressing whether dental chipping associated with grit in the diet (Towle, Irish, and De Groote 2017) can come from eating social insects directly from their nests.

Coprolites

Much of our knowledge of selected insect species for nonhuman primates comes from analysis of fecal matter. Chitinous insect parts often preserve through the digestive tract and remain diagnostic for determining insect species. Primate fecal matter can be collected even when the primates are not habituated to the observer's presence. It can take years of following a population before researchers can trust that what they are observing is the natural behavior of the primate and is not reactive behavior or behavior that is otherwise influenced by the presence of a human researcher. During this time, primatologists can use approaches similar to those archaeologists use to learn about their study population through analysis of what gets left behind, such as tools, nests, and, of course, fecal matter.[3]

Coprolites have the potential to greatly contribute to our knowledge of the extent of hominin uses of insects as food. Of course, this requires researchers to find the coprolite of a hominin that happened to eat insects the previous day. The probability is low, but it is not zero. The reassuring thing is that if a hominin coprolite is found, it can be examined in great detail, and if insect parts are present, their taxonomy can be identified.

When researchers encounter coprolites in archaeological contexts, they often assume that the presence of insects is post-depositional, that the insects inhabited the fecal matter after it became available in their environment. Unless the researcher has a particular interest in insects as food, he or she often does not identify these remains. This research gap is not surprising; archaeologists who encounter such evidence would need to collaborate with an entomologist to identify the insects and determine if they were consumed. The dietary bias of most westerners against insect consumption prevents many archaeologists from even thinking of insects as a possible food source, so those who ignore insects in coprolites likely do so inadvertently rather than intentionally.

Despite this common oversight, a lot has been learned about edible insects through analysis of coprolites. Much of what we know about the historic use of insects as food among indigenous populations of what is now the United States comes from this kind of evidence. Notable occurrences are from Danger Cave, Hogup Cave, and Glen Canyon, all in Utah; Dirty Shame Rockshelter in Oregon; Bamert Cave and Fish Slough Cave in California; Wetherill Mesa in Colorado; and Antelope House Pueblo and Mesa Verde in Arizona. Some of these sites are over 1,000 years old (Sutton 1995).

The likelihood that coprolites dating to 1,000 years ago have preserved and will be discovered is definitely higher than the likelihood that someone will discover a coprolite from a million years ago. In addition to the shorter time 1,000-year-old coprolites have been exposed to disturbance and degradation, archaeological sites that are more recent are easier to find. Recent habitation sites are easily identifiable by the presence of structures and by an abundance of artifacts. If people spent a lot of time there, there is a high probability that they had a place to relieve themselves nearby.

Because of hominin mobility, paleoanthropologists have a harder time finding habitation sites, but Neanderthals occupied caves, so we know more about their lives than we do about the lives of other species. In fact, the oldest human coprolite comes from a Neanderthal site. The site of El Salt in southern Spain contained a coprolite buried in sediment layers that date to about 50,000 years ago (Sistiaga et al. 2014). Unfortunately, the Neanderthal coprolite did not contain any insects. Because Neanderthals lived in northern environments, they would not have consumed many insects, and if and when they did, it would have been a seasonal and rare occurrence. If evidence of insect eating is to be found in a hominin coprolite, it likely will come from a site at a latitude that is much closer to the equator. Such a coprolite could belong to *Homo erectus* or an australopithecine, and it would be quite the discovery if it were to be found. The best chance of finding a coprolite is in an area of known habitation, which with these earlier hominins tends to be butchering sites, where stone artifacts and butchered bone occur in abundance. Will one ever be found? Only time will tell. It is encouraging that there are quite a few known dinosaur coprolites; knowing this makes finding one from a fossil hominin feel less impossible.

Fossil Insect Remains

Insects found in archaeological sediments are rarely studied. Most of the insects found at archaeological sites are undoubtedly invasive, and their remains are so small that they would not get caught in a standard sieve. Even if the remains were collected through use of a fine mesh screen, it would take countless hours to identify the taxa. Although insects may not be that interesting to the average archaeologist, their inclusion in the site is generally acknowledged. Archaeologists make painstaking efforts to ensure that all possible information is recorded so excavations can be reevaluated later, when new methods of analysis are developed. These methods of documentation, which include detailed records of the entire excavation process using notes and photographs, are why archaeology is a scientific field and differs from looting, or the collection of artifacts with no regard to their context. Collecting a soil sample from each excavation layer is also common practice in archaeological excavations. These samples preserve valuable data such as elements that are useful for determining the age of the site and plant phytoliths that can be used to reconstruct the environment and potentially also the available diet. Even the remains of insects are saved in these samples and they can be useful in archaeological analysis (Elias 2010).

Analyzing insect remains from soil samples is not an easy task, and it is generally not worth embarking upon unless the sediments come from a known occupation floor and other contextual information from the site suggests that there is value to the investigation. Ultimately the same information can be gained from this research that scientists gather for any fauna found at a site: total count, number of distinct taxa, minimum number of individuals, and frequencies of body elements. However, these data are most useful when they have been collected at several sites, contextualized, and compared. When comparative data is present, such as the record of data that is currently available for other fauna and flora, researchers can assess the ancient environment and identify potential food sources. Nonhuman remains from archaeological sites help researchers reconstruct an ecosystem, and any habitat would have included insect species. Revisiting collected sediments may reveal the utility of insects in reconstructing past environments and identifying potential foods that were available.

In order to identify insects used as food from their remains preserved in sediments, data first must be generated from a number of different contexts so that researchers can compare finds and identify deviances from expected patterns. To begin with, it is necessary to discern the frequency of the body parts of insects that occur naturally. With this information, we may be able to identify patterns that suggest the consumption of edible species. For instance, the wings of termite alates are commonly removed before eating them, as are the legs of grasshoppers. At a site where these insects were eaten, these discarded body elements should be found at higher frequencies in archaeological contexts.

Stable Isotope Analysis

Stable isotope analysis is becoming the norm in research on past diets. Both carbon and nitrogen preserve in the collagen of bone and can be analyzed to determine the proportions of different types of food that the individual consumed. Carbon is also preserved in dental enamel, which enables researchers to analyze stable carbon isotopes for more ancient hominins. However, the information such analysis provides is limited. The ratio of ^{12}C and ^{13}C isotopes reveals the proportions of the diet that come from two basic groups of plant foods, those that conduct photosynthesis using the C_3 pathway and those that use either the C_4 or the CAM pathway. However, the signature for eating a plant is the same as the signature for eating an animal that ate the same plant, and this limits the interpretive power of this analysis. When carbon analysis is combined with nitrogen analysis, a better estimate of the meat portion of the diet can be obtained. However, the bone collagen that contains nitrogen does not preserve in more ancient hominins. These data have been collected for Neanderthals, but it will be difficult to get them for specimens that are much older than Neanderthals.

In contrast to what happens with carbon, there is a positive shift in the nitrogen isotope with each trophic level. Thus, eating plants provides a small amount while consuming top-tier predators contributes much greater amounts. Certain patterns emerge when researchers look at both $\delta^{13}C$ and $\delta^{15}N$. Low amounts of both $\delta^{13}C$ and $\delta^{15}N$ suggest a vegetarian diet. Moderate amounts of $\delta^{15}N$ and low amounts of $\delta^{13}C$ suggest that an individual consumed terrestrial vertebrates. High amounts of both $\delta^{13}C$ and $\delta^{15}N$ suggest that an individual consumed marine fish. Maize is interesting because it is a domesticated grass, so finding high amounts of $\delta^{13}C$ and low amounts of $\delta^{15}N$ can suggest that an individual consumed this domesticate.

These different diet categories are very broad, they overlap with each other, and they are affected by environmental factors such as the amount of nitrogen in the soil. Additionally, there is no good way to parse out the proportions of foods included in a mixed diet when researchers find moderate amounts of both $\delta^{13}C$ and $\delta^{15}N$. How insects contribute to these isotopic signatures would depend on which insects were eaten, but even if that were known, the signature would not be precise enough to differ from other resource categories. For instance, isotopic analysis for the strepsirrhine primates known as bushbabies (*Galago* spp.), whose diet consists of fruit and insects, shows results that seem to suggest that they consume terrestrial vertebrates (Schoeninger, Iwaniec, and Nash 1998). Thus, there is not yet a way to differentiate isotope signatures between insects and other animal-based foods. Isotopic analyses are an excellent way to generate a baseline understanding of a particular diet and are essential when starting the analysis of any diet. Determination of specific foods, however, must come from other methods.

DNA

Genetic studies are becoming more sophisticated with each passing year. For instance, researchers are now able to reliably sequence DNA as ancient as that of Neanderthals and other similarly aged hominins. They can extract DNA from increasingly smaller samples and they can sequence the genomes of the bacteria that cohabitate with us (the microbiome). There is unlimited potential for genetic analysis to provide information about edible insects. Pickett and colleagues (2012) demonstrated that taxonomic identification of edible insect species is possible from insect DNA extracted from primate fecal matter. These methods could be used to identify fossil insect species from remains found in coprolites, sediments, and dental calculus. Research on the microbiomes of extant populations whose members consume insects also has the potential to determine which gut bacteria are associated with the insect consumption. Although this cannot be directly applied to our fossil hominin ancestors, comparative research across the great apes, including humans, may reveal the timing of the evolutionary origins of certain gut bacteria symbioses.

* * *

Direct evidence of insect consumption is limited in the archaeological record, but data from extant populations suggest that it was probably an important resource. In order to go forward in reconstructing this portion of the diet, standardization of data collection from extant populations is important for refining the models. Additionally, methods used to extract direct evidence from fossils and artifacts can be used to test hypotheses of insect consumption, but only if we first develop a comparative sample from experimental studies and analyses of present-day edible insects. Our biases against insects as food have left this gap in our knowledge, and in order for these studies to progress, anthropologists need to fully embrace the concept of insects as a food source. Overcoming this stigma and appreciating the value of edible insects is important for more accurately reconstructing hominin diets and for investigating the potential for the wider use of insects as food today.

9

Going Forward

Getting Over Our Obsession with Meat

The major themes of this book can be summarized as two principal points. The first theme is that insects are an excellent food option that our hominin ancestors likely exploited when they were available. The fact that some populations today avoid insects as food is a deviation that developed after thousands of years of human dispersal and complex cultural shaping. In modern society, we rarely do what is optimal for our environment or ourselves. Our complex brains and social organizations enable us to perform almost any behavior. However, we have abandoned the healthy behaviors our ancient ancestors practiced so they could survive and have replaced them with destructive ones that lead to all sorts of health problems. For instance, it is hard to imagine an obese early hominin, but today the rates of metabolic disease are alarming. Our industrialized food systems make poor-quality food options the ones that are most readily available, and it can take great effort and significant financial resources to supply our bodies with nourishing food. In light of the health issues that pervade today's industrialized societies, it is not surprising that many people are searching for sources of nourishment from outside the industrial food complex. When people opt for foods that are less processed, they are not only choosing healthful options but are also reducing their reliance on an industrial system that has flourished at the expense of our environment. Edible insects offer a possible alternative to commercially raised livestock, yielding similar nutritional benefits as meat while having a smaller impact on the environment. But insects are not a food option simply because of calculated nutritional and environmental benefits; they are also enjoyed by over two billion people around the today (Van Huis et al. 2013). It would be hasty to dismiss the potential insects have in the future of our modern diets.

The second theme of this book is a proposed hypothesis that posits that insects were an important food source targeted by female hominins over the course of human evolution. In addition to highlighting the benefits of insect consumption, this model draws attention to women, whose behavioral role is often ignored in paleoanthropology. One reason for their underrepresentation is that the field has been historically conducted by men so there is an inherent bias in the kinds of questions that long have been asked. Today, women scholars are well represented in paleoanthropology, but our training is still based on the traditions of the discipline. It often takes a concerted effort and some knowledge of feminist theory to identify these biases; just being a woman does not correct for centuries of one-sided arguments. Another way that the behavioral role of our women ancestors has been overlooked is that researchers have overemphasized the role of meat in the hominin diet. The archaeological record demonstrates that hominins butchered animals, and because of the well-known pattern in modern foragers of men doing most of the large-game hunting, the hominin males are central in researchers' reconstructions of resource acquisition. These reconstructions rely on the classic conceptualization of the cooperative provisioning model, where men are portrayed as hunting large game and bringing food home to provision their families. The underlying concept is that without the man's help, a woman would not be able to procure enough resources to support her large-bodied, large-brained baby (Brightman 1996). However, this goes against everything we see in the ethnographic record. Large game is not easy to catch, even for skillful hunters, and hunting parties often come home empty-handed. As Rebecca Bird states, "Many studies demonstrate that if men's subsistence strategies are simply about supplying energy to the household, they have plenty of room for improvement" (1999, 68). While men target high-risk resources, women in foraging societies tend to focus on resources that have a low risk of pursuit failure. These "gathered" foods are often plant-based, but foods such as invertebrates, small game, and eggs are also collected disproportionately more by women, and these foods often offer the same nutrients as large game. With a more critical view of how nutritional requirements are met in foraging societies, it becomes clear that the role of women is indispensable and far too important to be overlooked in reconstructions of hominin behavior.

Going forward, we must think beyond meat, both for the past and for the future. Paleoanthropological reconstructions suffer from a failure to acknowledge the breadth of dietary resources available to hominins. These meat-centric models perpetuate the unsustainable food economies that exist today.

Beyond Meat in Paleo Diets

The archaeological record of hominins consists mostly of stone tools. Although it is reasonable to think that hominins also used nonstone tools, such as sticks, these items are perishable and did not preserve in the archaeological record. Little, if any, attention is given to perishable items, and questions about behavior tend to center on the tools made of stone. The question then turns to, "What were these stone tools being used for?" Again, the answers are based on the evidence. Butchered animal bones are common at hominin sites. These bones preserve percussion marks, cut marks, and fracture patterns that are consistent with the use of stone tools to disarticulate limbs, remove flesh, and break open long bones to provide access to marrow. We do not have the same direct evidence of early hominin use of other food resources, at least not evidence that is as easily available as the evidence stone tools provide. Scholars do not think hominins were obligate carnivores, but because of the many analyses of stone tools and butchered animal bones, this behavior is what most vividly comes to mind when thinking about past ways of life.

There is great potential for future research that focuses on foods that do not come from large game, such as plant foods. Archaeologists are skilled at looking for plant residue on artifacts, but these methods and interpretations can be difficult to apply to pre-agricultural times. For instance, the residue of maize on a ceramic pot suggests that a population was agricultural or at least that it traded with a neighboring agricultural group. It is unlikely that the maize ended up accidentally in the sediments that line a vessel, and if it did, just the presence of maize in the locale is informative. Interpretations of this kind are far murkier in pre-agricultural sites, however. When the residue of a nondomesticated plant is found on the edge of a Paleolithic stone tool, healthy skepticism necessitates that researchers rule out contamination by plant debris in the soil. This exclusion is difficult because all available plant foods occurred naturally in pre-agricultural times. Because of this, paleoanthropologists largely ignore the possibility of plant foods in a culture's diet when they do not have specific evidence of such foods.

Of all of the hominins, Neanderthals ate the most meat. Meat is one of the most reliable high-quality resources that northern latitude environments have to offer. Meat is available year-round, unlike plant foods, which are more abundant seasonally. Neanderthals are often portrayed in a way that can make them appear as obligate carnivores. However, an essentially modern, human-like species of the genus *Homo* would have had similar nutritional requirements to ours. Some of these essential nutrients, such as the ever-important folate, had to come from plant sources. Without folate, cells cannot replicate correctly, leading to detrimental birth defects such as spina bifida. Neanderthals likely ate at least some leafy green plants that contained folate. Despite the logic of this science-based hypothesis, most available interpretations seem to point exclusively to meat eating.

Stable isotope analysis is becoming the new gold standard of dietary analysis for hominins, but we do not understand the processes behind the signatures as well as we would like to think. Nitrogen isotope signatures are used as a proxy for meat eating; values higher than $\delta^{15}N$ suggest more meat in the diet. Analysis of nitrogen for European Neanderthals suggests they have the highest $\delta^{15}N$ in their environments, higher even than European hyenas. This result confirmed for many that these hominins were carnivorous. However, Schoeninger (2014) has shown that a pattern in which human foragers are at the top of the trophic system also appears in other groups, such as archaic Great Basin foragers. These populations used simple hunting technologies but fairly sophisticated methods of gathering and processing foraged foods, including plants and insects (Madsen and Schmitt 1998). Although the Great Basin foragers were far from carnivorous, their nitrogen signatures are similar to that of Neanderthals and also higher than that of local canine species. Schoeninger notes the cause of the $\delta^{15}N$ spike is not clear for either group, but as there are no ethnographic accounts of obligate carnivory in any human group, it is likely that Neanderthals were eating more plant materials than we commonly think and that a yet-to-be-determined factor is skewing their isotope signature.

While it is difficult to provide convincing evidence of plant material in the hominin diet, Henry and colleagues (2011) have demonstrated that Neanderthals not only ate plants but also cooked them. The team analyzed plant microfossils, including the starch grains and phytoliths that were trapped in the dental calculus of two Neanderthal individuals from the site of Spy in Belgium and one Neanderthal individual from Shanidar in Iraq. These sites represent the extreme ends of the Neanderthal range, from one of the most eastern sites (Shanidar) to one of the most northern sites (Spy). At both sites, researchers found evidence of a wide spectrum of plant foods, including grasses, dates, legumes, the underground storage organs of plants, and other flora that researchers have yet to identify. Additionally, many of the starch grains had the distinct morphology that is the result of modification by cooking. Henry and colleagues note that the Neanderthals' consumption of these starchy plant foods does not contradict data from isotope analysis, because nitrogen isotopes record only the consumption of meat and protein-rich plant foods. Starchy plant foods would not contribute to the $\delta^{15}N$ signature.

Studies like these are important for considering elements of the hominin diet besides meat. It often takes this sort of direct evidence to convince paleoanthropologists of the utility of studying other food items, but it takes imagination to look for it. There is no telling what other resources might become apparent through these studies. It is my deepest hope that a termite leg will be found in dental calculus one day! Fortunately, paleoanthropology has a different line of evidence for termite foraging for earlier australopithecine hominins: the bone tools from South African sites where fossils of robust australopithecines have been found. Experiments have shown that the wear patterns on the ends of these tools are the best match for tools that were used to dig into termite mounds. The innovative thinking of Backwell and d'Errico is what led to this discovery (Backwell and d'Errico 2001). One of my goals for this book is to encourage paleoanthropology researchers to embrace creativity in their analysis of evidence of past diets because it will be these ideas that will continue to reveal underrepresented resources. It is highly probable that foods that today's researchers often consider marginal played a greater role in past diets than we have realized.

Another important aspect of moving past a meat-centric view of the hominin diet is the need to make continued efforts to include the role of women in reconstructions of past behavior. Numerous scholars have made efforts to note the importance of women to past societies, many of whom contributed to Dahlberg's 1981 edited volume titled *Woman the Gatherer*. The title of this book was a direct response to the 1968 volume *Man the Hunter*, edited by Lee and DeVore. Although Dahlberg's book is still in print, its reach and adoption have not matched that of *Man the Hunter*. A May 2017 search for the two books on Google Scholar showed that *Woman the Gatherer* has been cited 433 times, while *Man the Hunter* has been cited 1,925 times. The thirteen-year difference in publication dates accounts for some of the disparity, but the numbers also reflect how paleoanthropologists commonly dismiss female contributions in hominin life. Admittedly, it can be a tricky task to bring the role of women into the discussion, for fear of projecting our modern perceptions of gender onto the archaeological record. For instance, if a basket were to be found preserved in permafrost, it should not lead to the immediate interpretation that it was a woman who made or used it. However, we must make women more central in our scientific analysis of hominin behavior. If we continue to focus on meat procurement and hunting, we are leaving women out. Researchers have interpreted the phytoliths of woody plants found on Acheulean hand axes as a potential indicator that wood was being fashioned into spears (Dominguez-Rodrigo et al. 2001). Digging sticks just as easily could have been made, or branches could have been used to conceal a trap for small game, or strips of wood could have been used to make a basket. All of these speculations are valid, but the history of our field encourages us to consider the ones about men as more worthwhile. Broadening the range of conjecture makes women visible and subtly indicates that their roles were valuable.

Even in the literature on chimpanzee diets, meat has been known to govern interpretations. Chimpanzee males are noted hunters; they band together and go out as a party to catch prey, most commonly red colobus monkeys. A common interpretation of why female chimpanzees eat more insects than males is that they have to make up for the protein portion of the diet they are missing by not receiving the meat obtained during these hunts (McGrew 1979; Uehara 1986). However, when these hunts are successful, they provide meat that multiple individuals, even females, eat. A recent study by O'Malley and colleagues (2016) compared the volume of meat and insects female chimpanzees eat during pregnancy and early lactation to the volume of these foods they eat during nonreproductive phases. The dataset for this research, which spans thirty-eight years at Gombe National Park in Tanzania, revealed that higher-ranking females consumed more meat than lower-ranking females and that the higher-ranking females' meat consumption did not vary between reproductive states. However, lower-ranking females consumed more meat during pregnancy than lower-ranking females who were not pregnant or were lactating. For these females there was no difference in insect consumption based on rank or reproductive state. Therefore, meat consumption is mediated by social factors and reproductive state while insect consumption is consistent. Even when females have access to meat, they do not reduce the time they spend foraging for insects.

Although eating meat provides nutritional benefits for chimpanzees, complicated social networks are also at play. There may be benefits to avoiding meat, such as saving energy by not competing with others and preventing physical harm by avoiding the altercations that surround the distribution of meat. While a chimpanzee female can rise up through the ranks and earn the associated advantages, which includes access to meat, lower-ranking females still have opportunities to fulfill their daily nutritional needs by foraging for insects.

At Gombe, lower-ranking females have better access to meat while they are pregnant, which helps support their increased nutritional needs during that time. However, a different pattern is seen for orangutans, who have been observed eating meat only on rare occasions (Hardus et al. 2012). Using four years of behavioral data from Suaq Balimbing Research Station in Sumatra, Indonesia, Fox and colleagues (2004) found that pregnant females ate the most insects at their site, but the difference between their consumption and that by females with semi-dependent young was not significant. Interestingly, lactating females spent significantly less time feeding on insects and significantly more time feeding on fruit than females in other reproductive states, likely because manipulating tools high in the trees with dependent offspring in tow can create the risk of a potentially fatal fall for the infant. Female orangutans vary their insect consumption according to their needs. We should be celebrating the savvy of female apes, as well as women who forage for insects, because they commit to a reliable, nutritious resource. We need to recognize that meat is beneficial when it is available, but it is not necessarily critical.

Beyond Meat in the "Paleo Diet"

Paleoanthropologists who study diet are often quick to critique the popular "paleo diet" trend. This weight-loss program is predicated on eating the types of foods that people presume early humans ate, chiefly meat, fish, vegetables, and fruit, and excluding dairy or grain products and processed food (Cordain 2010) (figure 9.1). The strongest critique from anthropologists is that there is no singular way to describe what our ancestors were eating or even what modern foragers consume (Melnick 2014). For most of human evolution, our ancestors were in tropical regions, but even in the tropics, habitats can range from dry desert to lush forests. As *H. sapiens* spread around the globe, humans occupied diverse environments, including those at northern latitudes. What makes the human species successful is that we can make do in any of these places, as we do not rely exclusively on any single resource. Variety is central to the human diet. We get the energy and nutrients we need by eating a wide range of both plant and animal foods. The proportions of these resources vary by place, but meat is generally held in high esteem because it is high in quality and is often difficult to obtain.

Those who promote the paleo diet suggest that people today should eat more protein and more fat but reduce their intake of carbohydrates, especially from starchy foods. These guidelines usually result in a diet that is heavily flesh-centered. It is not surprising that "paleo" means "meat" to most people; paleoanthropological reconstructions of the diets of early hominins typically overemphasize meat. However, the paleo diet is purportedly based on models created from modern foragers, and this model does not hold up when we look at the ethnographic record. For example, the men of the Hadza in northern Tanzania hunt for large game, but tubers make up a larger portion of the diet. If starchy foods were to be restricted in this culture, as the paleo diet suggests, the Hadza would be missing one of their most critical food items. Hadza rank tubers very low in preference, but their continual availability make them an important fallback food when more preferred foods are unavailable (Marlowe and Berbesque 2009). This dietary flexibility exemplifies a true paleo diet. If we were to make our western diets more "paleo," we would emphasize seasonally available foods. That would include having days, maybe weeks, when meat is not consumed.

Figure 9.1. A paleo-diet-friendly kabob featuring grasshoppers that were wild harvested in Texas. Photo by C. M Cassady.

Eating large amounts of meat may have some people shedding pounds, but there is abundant evidence to suggest that reducing meat intake also has great benefits. Plant-based diets reduce the incidence of numerous health problems such as heart disease, stroke, cancer, diabetes, and obesity (Hu 2003; Saxe et al. 2006; Ferdowsian and Barnard 2009; Tuso et al. 2013). A plant-based diet is high in fruits, vegetables, and whole grains. The paleo diet similarly emphasizes these wholesome and nutritious foods, but most people would prefer to add an additional side of bacon to their plate instead of kale or broccoli, and paleo dieters can do so without cheating on their diets.

Another factor that many people today consider when choosing what to eat is the environmental impact of food. The meat industry uses nearly one-third of the world's arable land and about one-third of our fresh water, and it produces almost one-fifth of human-generated greenhouse gases (Steinfeld et al. 2006). Many people are concerned about how we can continue to eat as we do and still have the resources to feed the nine billion people who are projected to be on the planet by 2050 (Godfray et al. 2010). Thus, people are looking for ways to alter their diets to make them more sustainable. Foods labeled "organic" or "natural" target these sensibilities, but these terms are often marketing tactics and are not genuine indicators of the reduced environmental impact of the product. Many of the most devoted environmentalists choose a vegetarian or vegan diet so they will not contribute to the meat industry's exploitation of the environment.

Going meatless is not for everyone. It is often too costly to cut meat out completely in developed nations such as the United States. Meat is a nourishing food that is relatively cheap because of government subsidies, and when a person is on a budget, vegetarian options may not be practical. However, just like turning off the water while brushing your teeth or turning off the lights when leaving a room, every little bit can help if everyone does their part. This idea has led to a recent resurgence of meatless Mondays (e.g., the Meatless Monday website at http://www.meatlessmonday.com/). Meatless Mondays were part of a nationwide food-rationing effort during World War I in order to ensure that enough food was available to send to troops overseas and, after the war, to send to populations in need (Mullendore 1941). Today, meatless Mondays are an easy way for people to reduce their meat intake and begin each week with a positive reminder to do something good for one's health and for the planet. People who avoid meat one day a week find new foods to enjoy. The one thing we can say with certainty about the way our ancient ancestors ate is that they embraced variety in their diets.

Learning enough to make informed food choices can be overwhelming. It seems as if a good food choice would be one that is not processed, possibly something that our ancestors sought, a food that is wholesome and nourishing, adds variety to the diet, and has been produced sustainably. It is not easy to find a food that fits all of these categories. There is no lack of alternatives to meat on the market, from tofu to seitan, but producing these substitutes requires a lot of processing. This raises questions about how whole, wholesome, and sustainable they are.

According to one study, producing soy protein isolate, a common ingredient in plant-based meat alternatives, is less environmentally friendly than producing meat because of the large amounts of water and fossil fuel energy required to extract the protein isolate from soymeal (Berardy, Costello, and Seager 2015). However, not all soy products use isolates, so although good options are available, it falls on the consumer to identify them.

Fruits and vegetables provide the best answer, and plant-based diets have excellent benefits, but eliminating all animal foods is both culturally difficult for westerners and nutritionally insufficient. Supplements and enriched foods can easily provide vitamin B_{12} and calcium, but essential amino acids that are well represented in animal foods are hard to obtain in plant foods without careful consideration. Thus, limiting, rather than eliminating, consumption of animal-based foods and acquiring them from sustainable sources is the best way the average person can increase the sustainability of their diet while not compromising their nutrition. The difficulty comes with locating—and being able to afford—sustainably raised sources. It has been through pondering this dilemma that the picky eater in me finally came around to the potential of insects in the western diet.[1]

Not only are insects nutritious and widely consumed in other parts of the world, they can also be produced more sustainably than traditionally raised livestock. Food conversion ratios, which measure the efficiency with which the bodies of livestock convert feed into products for human consumption, are one way to compare livestock impact on the environment. In general, animals with smaller body sizes convert feed more efficiently and have lower food conversion ratios, although genetic selection, the engineering of animal feeds, and precision in the timing of slaughter

Figure 9.2. Crickets and cricket powder (30 grams of each) made from crickets grown in the United States for human consumption. Photo by J. Lesnik.

have enabled the meat industry to optimize output. But even compared to the small-bodied chickens, for which a powerful industry does research on best practices, insects are more efficient (figure 9.2). House crickets on a diet of poultry feed have an estimated food conversion ratio of 1.47 when calculating the fresh weight of whole crickets as the output; the ratio is 1.84 if we assume that only 80 percent of the biomass may be nutritionally available (researchers have yet to determine how much humans can extract from the chitinous bodies) (Lundy and Parella 2015). Broiler chickens have a conversion ratio of 3.52 for their edible mass, while eggs have a conversion ratio of 2.4 for their edible mass (Best 2011).

Another area where edible insects offer an environmentally friendly alternative to traditional livestock is their potential to reduce the emission of greenhouse gases (GHG). Anthropogenic greenhouse gases are of concern because of their contribution to increasing global temperatures, and the livestock industry is one of our biggest GHG producers. Oonincx and colleagues (2010) measured the output of carbon dioxide (CO_2), methane (CH_4), and nitrous oxide (N_2O) of five insect species. The results varied significantly across the insect species, but in most cases, when calculated as emissions per unit of body mass, the insect values were lower than those for cattle and comparable to or lower than those for pigs. These results suggest that even though insect GHG emissions are not likely to be worse than what the cattle industry produces, careful consideration should still be made when choosing insect species for mass production. Crickets and mealworms both produced less than the values given for traditional livestock for each GHG. This confirms that these insects are appropriate choices for large-scale production. Termites, however, are known to produce large amounts of methane (Zimmerman et al. 1982). Like cows, termites eat cellulose-dense foods that are digested by methane-producing bacteria and protozoa. Although naturally occurring termites are an excellent food source, factory farming them would be severely detrimental to the environment.

Other benefits of raising insect livestock include lower ethical concerns over their well-being. Insects are live animals, so harm comes to them when they are produced as food. However, compared to mammalian livestock, which have much more complex central nervous systems, this harm is arguably less. Additionally, while the reduction in quality of life for factory-farmed mammalian livestock is extreme, it is easier for farmers to maintain agreeable conditions for insects, which often seek dark, secretive places in nature. Because of their small size, these conditions can even be met in small, home-scale farms. Personal farms are available on the market that enable people to raise crickets and mealworms at home. These insects are particularly easy to cultivate, and the supplies needed to make their containers and to care for them are quite affordable. Eating insects farmed at home can provide autonomy and reduce our dependence on industrialized animal products, which may be key to combatting the effects of the increasing global population.

Of course, none of these benefits can improve the lives of people residing in the United States or other western nations if people do not accept the idea of insects as food. In these nations, there is a strong cultural bias against eating insects—so strong, in fact, that it triggers disgust, which is a difficult emotion to overcome. However, research on consumer acceptance has shown promising results. "Bug banquets" and other events aimed at introducing people to edible insects have become increasingly popular over recent years. These gatherings generally help people respond more positively when they encounter these foods again. People are most likely to try and to respond positively to insect-based foods that are in a form they are already familiar with, such as chocolate chip cookies. The market for protein bars made with insects is growing, for example. Continued research on consumer acceptance needs to focus on these early adopters so we can understand the most effective ways to help people overcome their fears and take advantage of all the benefits edible insects have to offer (Looy and Wood 2006; Sogari 2015; House 2016).

* * *

Although it is easy to critique, the paleo diet brings up important points that illustrate how problematic our modernized food system is. The concept of a paleo diet also introduces people to concepts of human evolution in an everyday context. Because they acknowledge that insects are a natural, healthful food and were likely consumed by our ancestors, paleo dieters are good allies in the insects-as-food movement. Edible insects are a true paleo food that deserves more recognition in academic pursuits and in our modern diets. If insects were to become popular as part of this current diet trend, they would have the potential for becoming a significant and long-lasting part of our future food system. The general problem with fads is that they come and go, but once the numerous dietary benefits of insects gain acceptance in western culture, they will be here to stay. In the not-too-distant future, edible insects will have their place in our restaurants, markets, and kitchens.

NOTES

Chapter 1. Introduction to Entomophagy Anthropology

1. Since 2005, the term hominin has been gradually replacing the word hominid, as it is meant to more accurately reflect the genetic relationship of humans and chimpanzees (Chimpanzee Sequencing and Analysis Consortium 2005).

2. It is commonly stated that there are an estimated 10 quintillion (10,000,000, 000,000,000,000) individual insects alive at any given time (Department of Systematic Biology, Entomology Section, National Museum of Natural History n.d.). Extrapolations from this estimate suggest that the biomass of insects would be seventy times that of the human population (Jackman 2012). However, in a classic article, Curtis W. Sabrosky (1953) discusses how difficult it is to estimate counts of insect species or individuals.

3. Water statistics for traditional livestock come from a statement by Van Huis et al. (2013), based on research they conducted for the FAO. The water statistic for crickets comes from Halloran et al. (2017), which is an updated value from what Van Huis et al. presented in the FAO review.

Chapter 4. Nutrition and Reproductive Ecology

1. Unless otherwise indicated, this chapter relies on Gibson (2005) for information about nutrition.

2. Osteoporosis is a major public health concern for postmenopausal women today. There is an increased incidence in this group because the sex hormones that regulate the cycle of bone remodeling decrease after menopause. Dietary calcium may not prevent bone loss for this population. Why and when human menopause evolved is a widely discussed puzzle of human life history. See Hawkes and Coxworth (2013) for a recent review.

3. The iron levels of women track with their reproductive state (Miller 2014). It has long been thought that women lose iron during menstruation, but this view is more controversial since the publication of contradictory results by Clancy, Nenko, and Jasienska (2006).

4. The widely noted differences in food procurement strategies seen between men and women in foraging societies today would most accurately be described as "gendered." However, when speaking about the potential evolutionary origins of this pattern,

evolutionary ecologists use the term "sexual division of labor," as they may be referring to patterns seen in birds or what is being reconstructed for early hominins. For the hominin lineage, it is difficult to identify when we should speak of society using the terms we use for people today. I choose to treat "gender" as a concept of modernity and reserve it for strategies seen in the past thousands of years, not in the past millions of years.

Chapter 6. Reconstructing the Role of Insects in the Diets of Early Hominins

1. At Swartkrans, Members 1 and 2 have yielded the remains of *Homo erectus* (Brain et al. 1988). "The absence of this taxon in Member 3, from where most of the bone tools derive, suggests, but does not prove, that these implements were used by *P. robustus*" (Backwell and d'Errico 2008, 2881). At Drimolen, as of the publication of Moggi-Cecchi and colleagues (2010) there were over eighty hominin specimens with less than a dozen attributable to *Homo* sp. Most of the remains are attributed to *P. robustus*, suggesting that it is most probable that they were the users of the bone tools.

2. Chimpanzees have been observed going after alates during swarming periods, but it is not common across populations and is usually only reported anecdotally. Kiyono-Fuse (2008) offers a good account.

Chapter 8. The Potential for Future Discovery: Testing Hypotheses of Edible Insects

1. I am currently working with biogeochemist Clayton Magill (Magill et al. 2016) and primatologist Robert O'Malley (O'Malley and Power 2014) on a project funded by the Leakey Foundation titled "An Evaluation of Termite-Associated Hydrocarbon Signatures as an Influence on Prey Selectivity and an Ecological Signal for Chimpanzees and Olduvai Hominins."

2. Storage in plastic bags can transfer plasticizers to archaeological residue (Heron and Evershed 1993). See Marreiros, Gibaja Bao, and Ferreira Bicho (2016) for full discussion of residue analysis.

3. Camera traps are also very useful for capturing behavior in nonhabituated primates. See Sanz, Morgan, and Gulick (2004) for termite tool use in the Goualougo Triangle, Republic of Congo. Also see the videos of the Goualougo Triangle Ape Project at https://www.youtube.com/channel/UChjMpWDfsAqjuS3hXwV7gPQ.

Chapter 9. Going Forward: Getting Over Our Obsession with Meat

1. I focus on the potential for the sustainable cultivation of edible insects here in the United States, but it is important to note that scaling up insect consumption raises its own concerns. 1) Although the focus here is on farming, increased interest in insects as food can lead to problems of overharvesting. This concern already exists for popular food insects like the mopane worm (Illgner and Nel 2000). 2) The low environmental impact of insect farming is dependent on what the insects are fed (as well as other practices of farm management). Simply feeding crickets with chicken feed may not yield the benefits we want to see (Lundy and Parrella 2015). And 3) Adopting insects into our diet does not erase our colonial history. As investors look to profit from a burgeoning industry, it is important to structure the process in a democratic way that includes and benefits the marginalized peoples who first crafted these practices (Müller et al. 2016).

REFERENCES

Abe, T., D. E. Bignell, and M. Higashi. 2000. *Termites: Evolution, sociality, symbioses, ecology*. Dordrecht: Kluwer Academic Publishers.

Allon, O., A. Pascual-Garrido, and V. Sommer. 2012. Army ant defensive behaviour and chimpanzee predation success: Field experiments in Nigeria. *Journal of Zoology* 288 (4): 237–244.

Antón, S. C. 2003. Natural history of *Homo erectus*. *American Journal of Physical Anthropology* 122 (46): 126–170.

Asfaw, B., T. White, O. Lovejoy, B. Latimer, S. Simpson, and G. Suwa. 1999. *Australopithecus garhi*: A new species of early hominid from Ethiopia. *Science* 284 (5414): 629–635.

Backwell, L., and F. d'Errico. 2001. Evidence of termite foraging by Swartkrans early hominids. *Proceedings of the National Academy of Sciences* 98 (4): 1358–1363.

———. 2005. The origin of bone tool technology and the identification of early hominid cultural traditions. In *From tools to symbols: From early hominids to modern humans*, ed. F. d'Errico and L. Backwell, 238–275. Johannesburg: Witwatersrand University Press.

———. 2008. Early hominid bone tools from Drimolen, South Africa. *Journal of Archaeological Science* 35 (11): 2880–2894.

Badrian, N., A. Badrian, and R. L. Susman. 1981. Preliminary observations on the feeding behaviour of *Pan paniscus* in the Lomako forest of central Zaire. *Primates* 22 (2): 173–181.

Balaresque, P., N. Poulet, S. Cussat-Blanc, P. Gerard, L. Quintana-Murci, E. Heyer, and M. A. Jobling. 2015. Y-chromosome descent clusters and male differential reproductive success: Young lineage expansions dominate Asian pastoral nomadic populations. *European Journal of Human Genetics* 10 (October 23): 1413–1422.

Balter, V., and L. Simon. 2006. Diet and behavior of the Saint-Cesaire Neanderthal inferred from biogeochemical data inversion. *Journal of Human Evolution* 51 (4): 329–338.

Banjo, A. D., O. A. Lawal, and E. A. Songonuga. 2006. The nutritional value of fourteen species of edible insects in southwestern Nigeria. *African Journal of Biotechnology* 5 (3): 298–301.

Bar-Yosef, O., and A. Belfer-Cohen. 2001. From Africa to Eurasia—early dispersals. *Quaternary International* 75 (1): 19–28.

Bartholomew, G., and J. Birdsell. 1953. Ecology and the protohominids. *American Anthropologist* 55 (4): 481–498.

Bassey, M., E. Etuk, M. Ibe, and B. Ndon. 2001. *Diplazium sammatii*: Athyraceae ("Nyama Idim"): Age-related nutritional and antinutritional analysis. *Plant Foods for Human Nutrition* 56 (1): 7–12.

Bellomo, R. V. 1994. Methods of determining early hominid behavioral activities associated with the controlled use of fire at FxJj 20 Main, Koobi Fora, Kenva. *Journal of Human Evolution* 27 (1–3): 173–195.

Bentley, G. R. 1985. Hunter-gatherer energetics and fertility: A reassessment of the !Kung San. *Human Ecology* 13 (1): 79–109.

Berardy, A., C. Costello, and T. Seager. 2015. Life cycle assessment of soy protein isolate. *Proceedings of the International Symposium on Sustainable Systems and Technologies* 3. Accessed November 28, 2017. https://doi.org/10.6084/m9.figshare.1517821.v1.

Best, P. 2011. Poultry performance improves over past decades. *WATT AgNet.com*, November 24. Accessed November 28, 2017. https://www.wattagnet.com/articles/10427-poultry-performance-improves-over-past-decades.

Bird, R. 1999. Cooperation and conflict: The behavioral ecology of the sexual division of labor. *Evolutionary Anthropology: Issues, News, and Reviews* 8 (2): 65–75.

Bird, R., B. Codding, and D. Bird. 2009. What explains differences in men's and women's production? *Human Nature* 20 (2): 105–129.

Blake, E. A., and M. R. Wagner. 1987. Collection and consumption of pandora moth, Coloradia pandora lindseyi (Lepidoptera: Saturniidae), larvae by Owens Valley and Mono Lake Paiutes. *Bulletin of the ESA* 33 (1): 22–27.

Bloch, J. I., and M. T. Silcox. 2006. Cranial anatomy of the Paleocene plesiadapiform *Carpolestes simpsoni* (Mammalia, Primates) using ultra high-resolution X-ray computed tomography, and the relationships of plesiadapiforms to Euprimates. *Journal of Human Evolution* 50 (1):1–35.

Bocherens, H., D. Drucker, D. Billiou, M. Patou-Mathis, and B. Vandermeersch. 2005. Isotopic evidence for diet and subsistence pattern of the Saint-Césaire I Neanderthal: Review and use of a multi-source mixing model. *Journal of Human Evolution* 49 (1): 71–87.

Bodenheimer, F. S. 1951. *Insects as human food: A chapter of the ecology of man*. The Hague: Junk.

Bogart, S., and J. Pruetz. 2008. Ecological context of savanna chimpanzee (*Pan troglodytes verus*) termite fishing at Fongoli, Senegal. *American Journal of Primatology* 70 (6): 605–612.

Bogin, B. 2011. !Kung nutritional status and the original "affluent society"—a new analysis. *Anthropologischer Anzeiger* 68 (4): 349–366.

Boinski, S. 1988. Sex differences in the foraging behavior of squirrel monkeys in a seasonal habitat. *Behavioral Ecology and Sociobiology* 23 (3): 177–186.

Boyero, L. 2012. Latitudinal gradients in biodiversity. *Ecology.Info* 32. Accessed November 28, 2017. http://www.ecology.info/gradients-biodiversity.htm.

Brain, C. K. 1981. *The hunters or the hunted? An introduction to African cave taphonomy*. Chicago: University of Chicago Press.

——, ed. 1993. *Swartkrans: A cave's chronicle of early man*, 2nd ed. Pretoria: Transvaal Museum.

Brain, C. K., C. S. Churcher, J. D. Clark, F. E. Grine, P. Shipman, R. L. Susman, A. Turner, and V. Watson. 1988. New evidence of early hominids, their culture, and environment from the Swartkrans cave. *South African Journal of Science* 84 (10): 828–835.

Brain, C. K., and P. Shipman. 1993. The Swartkrans bone tools. In *Swartkrans: A cave's chronicle of early man*, 2nd ed., ed. C. K. Brain, 192–215. Pretoria: Transvaal Museum.

Brandl, R., F. Hyodo, M. von Korff-Schmising, K. Maekawa, T. Miura, Y. Takematsu, Y. Matsumoto, T. Abe, R. Bagine, and M. Kaib. 2007. Divergence times in the termite genus *Macrotermes* (Isoptera: Termitidae). *Molecular phylogenetics and evolution* 45 (1): 239–250.

Breslin, P. A. 2013. An evolutionary perspective on food and human taste. *Current Biology* 23 (9): R409–R418.

Brightman, R. 1996. The sexual division of foraging labor: Biology, taboo, and gender politics. *Comparative Studies in Society and History* 38 (4): 687–729.

Brown, J. K. 1970. A note on the division of labor by sex. *American Anthropologist* 72 (5): 1073–1078.

Buck, L. T., and C. B. Stringer. 2014. Having the stomach for it: A contribution to Neanderthal diets? *Quaternary Science Reviews* 96: 161–167.

Bugsolutely. 2018. List of edible insect companies. Accessed January 9, 2018. http://www.bugsolutely.com/yellow-bug-pages/.

Bukkens, S. G. F. 1997. The nutritional value of edible insects. *Ecology of Food and Nutrition* 36 (2–4): 287–319.

Bunn, H. T. 1981. Archaeological evidence for meat-eating by Plio-Pleistocene hominids from Koobi Fora and Olduvai Gorge. *Nature* 291: 574–577.

Byrne R. W., P. J. Barnard, I. Davidson, V. M. Janik, W. C. McGrew, A. Miklósi, and P. Wiessner. 2004. Understanding culture across species. *Trends in Cognitive Sciences* 8: 341–346.

Call, J., and M. Tomasello. 1994. The social learning of tool use by orangutans (*Pongo pygmaeus*). *Human Evolution* 9 (4): 297–313.

Callen, E. O. 1965. Food habits of some pre-Columbian Mexican Indians. *Economic Botany* 19 (4): 335–343.

Caparros Megido, R., L. Sablon, M. Geuens, Y. Brostaux, T. Alabi, C. Blecker, D. Drugmand, E. Haubruge, and F. Francis. 2014. Edible insects acceptance by Belgian consumers: Promising attitude for entomophagy development. *Journal of Sensory Studies* 29 (1): 14–20.

Carew, J. 1988a. Columbus and the origins of racism in the Americas: Part one. *Race & Class* 29 (4): 171–176.

———. 1988b. Columbus and the origins of racism in the Americas: Part two. *Race & Class* 30 (1): 33–57.

Carmody, R. N., and R. W. Wrangham. 2009. The energetic significance of cooking. *Journal of Human Evolution* 57 (4): 379–391.

Carr, L. G. 1951. Interesting animal foods, medicines, and omens of the eastern Indians, with comparisons to ancient European practices. *Journal of the Washington Academy of Sciences* 41 (7): 229–235.

Cartmill, M. 1974. Rethinking primate origins. *Science* 184 (4135): 436–443.

Cerling, T. E., F. K. Manthi, E. N. Mbua, L. N. Leakey, M. G. Leakey, R. E. Leakey, F. H. Brown, F. E. Grine, J. A. Hart, P. Kaleme, H. Roche, K. T. Uno, and B. A. Wood. 2013. Stable isotope-based diet reconstructions of Turkana Basin hominins. *Proceedings of the National Academy of Sciences* 110 (26): 10501–10506.

Cerling, T. E., E. Mbua, F. M. Kirera, F. K. Manthi, F. E. Grine, M. G. Leakey, M. Sponheimer, and K. T. Uno. 2011. Diet of *Paranthropus boisei* in the early Pleistocene of East Africa. *Proceedings of the National Academy of Sciences* 108 (23): 9337–9341.

Cerritos Flores, R., R. Ponce-Reyes, and F. Rojas-García. 2014. Exploiting a pest insect species *Sphenarium purpurascens* for human consumption: Ecological, social, and economic repercussions. *Journal of Insects as Food and Feed* 1 (1): 75–84.

Chakravorty, J., S. Ghosh, and V. B. Meyer-Rochow. 2011. Practices of entomophagy and entomotherapy by members of the Nyishi and Galo tribes, two ethnic groups of the state of Arunachal Pradesh (North-East India). *Journal of Ethnobiology and Ethnomedicine* 7 (1): 5–5.

Charnov, E. L. 1976. Optimal foraging, the marginal value theorem. *Theoretical Population Biology* 9 (2): 129–136.

Cherry, R. H. 1991. Use of insects by Australian aborigines. *American Entomologist* 37 (1): 8–13.

Chiltoskey, M. U. 1975. Cherokee Indian foods. In *Gastronomy: The anthropology of food and food habits*, ed. M. L. Arnott, 235–244. Chicago: Aldine Publishing Company.

Chimpanzee Sequencing and Analysis Consortium. 2005. Initial sequence of the chimpanzee genome and comparison with the human genome. *Nature* 437 (7055): 69–87.

Cipolletta, C., N. Spagnoletti, A. Todd, M. M. Robbins, H. Cohen, and S. Pacyna. 2007. Termite feeding by *Gorilla gorilla gorilla* at Bai Hokou, Central African Republic. *International Journal of Primatology* 28 (2): 457–476.

Clancy, K. B., I. Nenko, and G. Jasienska. 2006. Menstruation does not cause anemia: Endometrial thickness correlates positively with erythrocyte count and hemoglobin concentration in premenopausal women. *American Journal of Human Biology* 18 (5): 710–713.

Codron, D., J. Lee-Thorp, M. Sponheimer, and J. Codron. 2007. Nutritional content of savanna plant foods: Implications for browser/grazer models of ungulate diversification. *European Journal of Wildlife Research* 53 (2): 100–111.

Codron, D., J. Lee-Thorp, M. Sponheimer, D. de Ruiter, and J. Codron. 2006. Inter- and intrahabitat dietary variability of chacma baboons (*Papio ursinus*) in South African savannas based on fecal $\delta^{13}C$, $\delta^{15}N$, and %N. *American Journal of Physical Anthropology* 129 (2): 204–214.

Cohen, J. H., N. D. M. Sánchez, and F. Montiel-Ishino. 2009. *Chapulines* and food choices in rural Oaxaca. *Gastronomica* 9 (1): 61–65.

Collins, D. A., and W. C. McGrew. 1985. Chimpanzees' (*Pan troglodytes*) choice of prey among termites (Macrotermitinae) in western Tanzania. *Primates* 26 (4): 375–389.

Conklin-Brittain, N. L., R. W. Wrangham, and K. D. Hunt. 1998. Dietary response of chimpanzees and cercopithecines to seasonal variation in fruit abundance. II. Macronutrients. *International Journal of Primatology* 19 (6): 971–998.

Cordain, L. 2010. *The paleo diet revised: Lose weight and get healthy by eating the foods you were designed to eat.* Hoboken: John Wiley and Sons, Inc.

Cortés-Sánchez, M., A. Morales-Muñiz, M. D. Simón-Vallejo, M. C. Lozano-Francisco, J. L. Vera-Peláez, C. Finlayson, J. Rodríguez-Vidal, A. Delgado-Huertas, F. J. Jiménez-Espejo, F. Martínez-Ruiz, and M. A. Martínez-Aguirre. 2011. Earliest known use of marine resources by Neanderthals. *PLoS ONE* 6 (9): e24026.

Crowley, P. 2014a. Can eating insects solve global problems in an ever-changing world? Paper presented at TedXJacksonHole, Jackson Hole, Wyoming. Accessed January 9, 2018. https://www.youtube.com/watch?v=JNks0CQdzT8.

———. 2014b. Why not eat insects? Paper presented at the TEDxJacksonHole. YouTube video. Accessed November 28, 2017. https://www.youtube.com/watch?v=uzrpSXqp40g.

———. 2015. Eating insects. Paper presented at TedXZwolle, Zwolle, Netherlands. Accessed January 9, 2018. https://www.youtube.com/watch?v=gvX7kVUhrkw.

Curtis, V., R. Aunger, and T. Rabie. 2004. Evidence that disgust evolved to protect from risk of disease. *Proceedings of the Royal Society of London B: Biological Sciences* 271 (Suppl. 4): S131–S133.

d'Errico, F., and L. R. Backwell. 2003. Possible evidence of bone tool shaping by Swartkrans early hominids. *Journal of Archaeological Science* 30 (12): 1559–1576.

———. 2009. Assessing the function of early hominin bone tools. *Journal of Archaeological Science* 36 (8): 1764–1773.

d'Errico, F., C. Henshilwood, M. Vanhaeren, and K. Van Niekerk. 2005. *Nassarius kraussianus* shell beads from Blombos Cave: Evidence for symbolic behaviour in the Middle Stone Age. *Journal of Human Evolution* 48 (1): 3–24.

Dahlberg, F. 1981. *Woman the gatherer.* New Haven, CT: Yale University Press.

Dart, R. A. 1925. *Australopithecus africanus*: The man-ape of South Africa. *Nature* 115 (February 7): 195–199.

———. 1957. *The osteodontokeratic culture of* Australopithecus prometheus. Pretoria: Transvaal Museum.

Davey, G. C. 1994. The "disgusting" spider: The role of disease and illness in the perpetuation of fear of spiders. *Society and Animals* 2 (1): 17–25.

de Pablo, J. F.-L., E. Badal, C. F. García, A. Martínez-Ortí, and A. S. Serra. 2014. Land snails as a diet diversification proxy during the early upper Palaeolithic in Europe. *PLoS ONE* 9 (8): e104898.

Deblauwe, I., J. Dupain, G. M. Nguenang, D. Werdenich, and L. Van Elsacker. 2003. Insectivory by *Gorilla gorilla gorilla* in southeast Cameroon. *International Journal of Primatology* 24 (3): 493–502.

Deblauwe, I., and G. P. Janssens. 2008. New insights in insect prey choice by chimpanzees and gorillas in southeast Cameroon: The role of nutritional value. *American Journal of Physical Anthropology* 135 (1): 42–55.

DeFoliart, G. R. 1999. Insects as food: Why the western attitude is important. *Annual Review of Entomology* 44 (1): 21–50.

———. 2002. *The human use of insects as a food resource: A bibliographic account in progress*. Madison: University of Wisconsin.

DeLoache, J. S., and V. LoBue. 2009. The narrow fellow in the grass: Human infants associate snakes and fear. *Developmental Science* 12 (1): 201–207.

Department of Systematic Biology, Entomology Section, National Museum of Natural History. N.d. BugInfo: Fun facts about bugs. *Smithsonian*. Accessed February 8, 2018. https://www.si.edu/spotlight/buginfo.

DeSilva, J. M., and J. J. Lesnik. 2008. Brain size at birth throughout human evolution: A new method for estimating neonatal brain size in hominins. *Journal of Human Evolution* 55 (6): 1064–1074.

Dicke, M. 2010. Why not eat insects? Presentation at TEDGlobal, London, England. Accessed November 28, 2017. http://www.ted.com/talks/marcel_dicke_why_not_eat_insects?language=en.

Dillehay, T. D., C. Ocampo, J. Saavedra, A. O. Sawakuchi, R. M. Vega, M. Pino, M. B. Collins, L. S. Collins, I. Arregui, X. S. Villagran, G. A. Hartmann, M. Mella, A. González, and G. Dix. 2015. New archaeological evidence for an early human presence at Monte Verde, Chile. *PLoS ONE* 10 (12): e0145471.

Dominguez-Rodrigo, M., T. R. Pickering, S. Semaw, and M. J. Rogers. 2005. Cutmarked bones from Pliocene archaeological sites at Gona, Afar, Ethiopia: Implications for the function of the world's oldest stone tools. *Journal of Human Evolution* 48 (2): 109–121.

Dominguez-Rodrigo, M., J. Serrallonga, J. Juan-Tresserras, L. Alcala, and L. Luque. 2001. Woodworking activities by early humans: A plant residue analysis on Acheulian stone tools from Peninj (Tanzania). *Journal of Human Evolution* 40 (4): 289–299.

Doran-Sheehy, D., P. Mongo, J. Lodwick, and N. L. Conklin-Brittain. 2009. Male and female western gorilla diet: Preferred foods, use of fallback resources, and implications for ape versus Old World monkey foraging strategies. *American Journal of Physical Anthropology* 140 (4): 727–738.

Dufour, D. L. 1987. Insects as food: A case study from the northwest Amazon. *American Anthropologist* 89 (2): 383–397.

Dusseldorp, G. L. 2011. Studying Pleistocene Neanderthal and cave hyena dietary habits: Combining isotopic and archaeozoological analyses. *Journal of Archaeological Method and Theory* 18 (3): 224–255.

Earle, R. 2010. "If you eat their food . . .": Diets and bodies in early colonial Spanish America. *American Historical Review* 115 (3): 688–713.

Elias, S. A. 2010. The use of insect fossils in archeology. *Developments in Quaternary Sciences* 12: 89–121.

Evans, J., M. H. A. Alemu, R. Flore, M. B. Frøst, A. Halloran, A. B. Jensen, G. Maciel-Vergara, V. B. Meyer-Rochow, C. Münke-Svendsen, S. B. Olsen, C. Payne, N. Roos, P. Rozin, H. S. G. Tan, A. Van Huis, P. Vantomme, and J. Eilenberg. 2015. "Entomophagy": An evolving terminology in need of review. *Journal of Insects as Food and Feed* 1 (4): 293–305.

Evershed, R. P. 2008. Organic residue analysis in archaeology: The archaeological biomarker revolution. *Archaeometry* 50 (6): 895–924.

Felt, E. P. 1918. Caribou warble grubs edible. *Journal of Economic Entomology* 11: 482.

Ferdowsian, H. R., and N. D. Barnard. 2009. Effects of plant-based diets on plasma lipids. *American Journal of Cardiology* 104 (7): 947–956.

Flood, J. 1980. *The moth hunters: Aboriginal prehistory of the Australian Alps*. Canberra, Australia: Australian Institute of Aboriginal Studies.

Food and Nutrition Board, Institute of Medicine. 2011. *Dietary reference intakes for Vitamin A, Vitamin K, arsenic, boron, chromium, copper, iodine, iron, manganese, molybdenum, nickel, silicon, vanadium, and zinc*. Washington, DC: National Academy Press.

Fox, E. A., C. P. Van Schaik, A. Sitompul, and D. N. Wright. 2004. Intra- and interpopulational differences in orangutan (*Pongo pygmaeus*) activity and diet: Implications for the invention of tool use. *American Journal of Physical Anthropology* 125 (2): 162–174.

Fry, G. F. 1978. Prehistoric diet at Danger Cave, Utah, as determined by analysis of coprolites. *University of Utah Anthropological Papers* 99 (23): 107–126.

Futuyma, D. J. 1998. *Evolutionary biology*. 3rd ed. Sunderland, MA: Sinauer Associates, Inc.

Galdikas, B. M. F. 1988. Orangutan diet, range, and activity at Tanjung Puting, Central Borneo. *International Journal of Primatology* 9 (1): 1–35.

Ganas, J., and M. M. Robbins. 2004. Intrapopulation differences in ant eating in the mountain gorillas of Bwindi Impenetrable National Park, Uganda. *Primates* 45 (4): 275–278.

Gandhi, M. 1954. *How to serve the cow*. Ahmedabab: Navajivan Publishing House.

Garber, P. A. 1987. Foraging strategies among living primates. *Annual Review of Anthropology* 16: 339–364.

Gaston, K. J. 2007. Latitudinal gradient in species richness. *Current Biology* 17 (15): R574.

Gerdes, A. B. M., G. Uhl, and G. W. Alpers. 2009. Spiders are special: Fear and disgust evoked by pictures of arthropods. *Evolution and Human Behavior* 30 (1): 66–73.

Gibson, R. S. 2005. *Principles of nutritional assessment*. 2nd ed. Oxford: Oxford University Press.

Godfray, H. C. J., J. R. Beddington, I. R. Crute, L. Haddad, D. Lawrence, J. F. Muir, J. Pretty, S. Robinson, S. M. Thomas, and C. Toulmin. 2010. Food security: The challenge of feeding 9 billion people. *Science* 327 (5967): 812–818.

Goertzen, C. 2010. *Made in Mexico: Tradition, tourism, and political ferment in Oaxaca*. Jackson: University Press of Mississippi.

Goodall, J. 1963. Feeding behaviour of wild chimpanzees: A preliminary report. *Symposium of the Zoological Society of London* 10: 39–47.

Goodall, J. 1998. Essays on science and society: Learning from the chimpanzees: A message humans can understand. *Science* 282 (5397): 2184–2185.

Goodenough, W. H. 1957. Cultural anthropology and linguistics. In *Report of the seventh annual round table meeting on linguistics and language study*, ed. P. I. Garvin, 167–173. Washington, DC: Georgetown University Press.

Gordon, D. G. 1998. *The eat-a-bug cookbook*. Berkeley: Ten Speed Press.

Goren-Inbar, N., N. Alperson, M. E. Kislev, O. Simchoni, Y. Melamed, A. Ben-Nun, and E. Werker. 2004. Evidence of hominin control of fire at Gesher Benot Ya'aqov, Israel. *Science* 304 (5671): 725–727.

Grine, F. E. 1986. Dental evidence for dietary differences in *Australopithecus* and *Paranthropus*: A quantitative analysis of permanent molar microwear. *Journal of Human Evolution* 15 (8): 783–822.

Gurven, M., and C. von Rueden. 2006. Hunting, social status and biological fitness. *Social Biology* 53 (1–2): 81–99.

Guthrie, R. D. 2005. *The nature of Paleolithic art*. Chicago: University of Chicago Press.

Hagenau, T., R. Vest, T. Gissel, C. Poulsen, M. Erlandsen, L. Mosekilde, and P. Vestergaard. 2009. Global vitamin D levels in relation to age, gender, skin pigmentation and latitude: An ecologic meta-regression analysis. *Osteoporosis International* 20 (1): 133–140.

Hall, H. J. 1977. Paleoscatological study of diet and disease at Dirty Shame Rockshelter, Southeast Oregon. *Tebiwa* 8: 1–15.

Halloran, A., Y. Hanboonsong, N. Roos, and S. Bruun. 2017. Life cycle assessment of cricket farming in north-eastern Thailand. *Journal of Cleaner Production* 156: 83–94.

Hardus, M. E., A. R. Lameira, A. Zulfa, S. S. U. Atmoko, H. de Vries, and S. A. Wich. 2012. Behavioral, ecological, and evolutionary aspects of meat-eating by Sumatran orangutans (*Pongo abelii*). *International Journal of Primatology* 33 (2): 287–304.

Harmand, S., J. E. Lewis, C. S. Feibel, C. J. Lepre, S. Prat, A. Lenoble, X. Boës, R. L. Quinn, M. Brenet, A. Arroyo, N. Taylor, S. Clément, G. Daver, J.-P. Brugal, L. Leakey, R. A. Mortlock, J. D. Wright, S. Lokorodi, C. Kirwa, D. V. Kent, and H. Roche. 2015. 3.3-million-year-old stone tools from Lomekwi 3, West Turkana, Kenya. *Nature* 521 (7552): 310–315.

Harris, M. 1966. *The cultural ecology of India's sacred cattle*. New York: Columbia University Department of Anthropology.

———. 1985. *Good to eat: Riddles of food and culture*. Long Grove, IL: Waveland Press.

Harris, M., N. K. Bose, M. Klass, J. P. Mencher, K. Oberg, M. K. Opler, W. Suttles, and A. P. Vayda. 1966. The cultural ecology of India's sacred cattle [and comments and replies]. *Current Anthropology* 7 (1): 51–66.

Hartmann, C., J. Shi, A. Giusto, and M. Siegrist. 2015. The psychology of eating insects: A cross-cultural comparison between Germany and China. *Food Quality and Preference* 44: 148–156.

Hawkes, K., and J. E. Coxworth. 2013. Grandmothers and the evolution of human longevity: A review of findings and future directions. *Evolutionary Anthropology* 22 (6): 294–302.

Hawkes, K., K. Hill, and J. F. O'Connell. 1982. Why hunters gather: Optimal foraging and the Aché of eastern Paraguay. *American Ethnologist* 9 (2): 379–398.

Hawkes, K., J. F. O'Connell, and N. G. B. Jones. 2001. Hadza meat sharing. *Evolution and Human Behavior* 22 (2): 113–142.

Hawks, J., S. Oh, K. Hunley, S. Dobson, G. Cabana, P. Dayalu, and M. H. Wolpoff. 2000. An Australasian test of the recent African origin theory using the WLH-50 calvarium. *Journal of Human Evolution* 39 (1): 1–22.

Hedges, S., and W. C. McGrew. 2012. Ecological aspects of chimpanzee insectivory in the Budongo Forest, Uganda. *Pan Africa News* 19: 6–7.

Henry, A. G., A. S. Brooks, and D. R. Piperno. 2011. Microfossils in calculus demonstrate consumption of plants and cooked foods in Neanderthal diets (Shanidar III, Iraq; Spy I and II, Belgium). *Proceedings of the National Academy of Sciences* 108 (2): 486–491.

Henshilwood, C. S., J. C. Sealy, R. Yates, K. Cruz-Uribe, P. Goldberg, F. E. Grine, R. G. Klein, C. Poggenpoel, K. van Niekerk, and I. Watts. 2001. Blombos Cave, Southern Cape, South Africa: Preliminary report on the 1992–1999 excavations of the Middle Stone Age levels. *Journal of Archaeological Science* 28 (4): 421–448.

Heron, C., and R. P. Evershed. 1993. The analysis of organic residues and the study of pottery use. *Archaeological Method and Theory* 5: 247–284.

Hewitt, G. M. 1999. Post-glacial re-colonization of European biota. *Biological Journal of the Linnean Society* 68 (1–2): 87–112.

Hill, K., K. Hawkes, M. Hurtado, and H. Kaplan. 1984. Seasonal variance in the diet of Aché hunter-gatherers in eastern Paraguay. *Human Ecology* 12 (2): 101–135.

Hill, K., and A. M. Hurtado. 1996. *Aché life history: The ecology and demography of a foraging people*. New York: Routledge.

Hiraiwa-Hasegawa, M. 1989. Sex differences in the behavioral development of chimpanzees at Mahale. In *Understanding Chimpanzees*, ed. P. G. Heltne and L. A. Marquardt, 104–111. Cambridge, MA: Harvard University.

Hladik, C. M. 1973. Alimentation et activité d'un groupe de chimpanzés réintroduits en forêt gabonaise. *La Terre et la vie* 27: 343–413.

Holloway, R. L., D. C. Broadfield, and M. S. Yuan. 2004. *The human fossil record*. Vol. 3, *Brain endocasts—The paleoneurological evidence*. Hoboken, NJ: Wiley-Liss.

Holt, V. M. 1885. *Why not eat insects?* London: Field and Tuer.

House, J. 2016. Consumer acceptance of insect-based foods in the Netherlands: Academic and commercial implications. *Appetite* 107: 47–58.

Hrdy, S. B. 1981. *The woman that never evolved*. Cambridge: Harvard University Press

Hu, F. B. 2003. Plant-based foods and prevention of cardiovascular disease: An overview. *American Journal of Clinical Nutrition* 78 (3): 544S–551S.

Hublin, J.-J., D. Weston, P. Gunz, M. Richards, W. Roebroeks, J. Glimmerveen, and L. Anthonis. 2009. Out of the North Sea: The Zeeland Ridges Neandertal. *Journal of Human Evolution* 57 (6): 777–785.

Humle, T., C. T. Snowdon, and T. Matsuzawa. 2009. Social influences on ant-dipping acquisition in the wild chimpanzees (*Pan troglodytes verus*) of Bossou, Guinea, West Africa. *Animal Cognition* 12 (1): 37–48.

Hurtado, A. M., K. Hawkes, K. Hill, and H. Kaplan. 1985. Female subsistence strategies among Aché hunter-gatherers of eastern Paraguay. *Human Ecology* 13 (1): 1–28.

Ichikawa, M. 1987. Food restrictions of the Mbuti pygmies, eastern Zaire. *African Study Monographs* 6: 97–121.

Illgner, P., and E. Nel. 2000. The geography of edible insects in Sub-Saharan Africa: A study of the mopane caterpillar. *The Geographical Journal* 166 (4): 336–351.

Institute of Medicine of the National Academies. 2005. *Dietary reference intakes for energy, carbohydrate, fiber, fat, fatty acids, cholesterol, protein, and amino acids (macronutrients)*. Washington, DC: National Academy Press.

———. 2010. *Dietary reference intakes: RDA and AI for vitamins and elements*. Washington, DC: National Academy Press.

Isbell, L. A. 1998. Diet for a small primate: Insectivory and gummivory in the (large) patas monkey (*Erythrocebus patas pyrrhonotus*). *American Journal of Primatology* 45 (4): 381–398.

———. 2006. Snakes as agents of evolutionary change in primate brains. *Journal of Human Evolution* 51 (1): 1–35.

Jablonski, N. G. 2013. *Skin: A natural history.* Berkeley: University of California Press.

Jackman, P. 2012. The weight of all those creepy crawlies. *Global Mail,* August 3. Accessed January 9, 2018. https://www.theglobeandmail.com/opinion/the-weight-of-all-those-creepy-crawlies/article4461850/.

Janiak, M. C., Chaney, M. E., and Tosi, A. J. 2017. Evolution of acidic mammalian chitinase genes (CHIA) is related to body mass and insectivory in primates. *Molecular Biology and Evolution* 35 (3): 607–622.

Janiszewski, P. 2015. How long can humans survive without food or water? *PLoS Blogs,* November 30, 2015. Accessed November 28, 2017. http://blogs.plos.org/obesitypanacea/2015/11/30/how-long-can-humans-survive-without-food-or-water/.

Jongema, Y. 2017. Worldwide list of edible insects. Accessed November 28, 2017. https://www.wur.nl/upload_mm/8/a/6/0fdfc700-3929-4a74-8b69-f02fd35a1696_Worldwide%20list%20of%20edible%20insects%202017.pdf.

Kaib, M., R. Brandl, and R. Bagine. 1991. Cuticular hydrocarbon profiles: A valuable tool in termite taxonomy. *Naturwissenschaften* 78 (4): 176–179.

Kappelman, J. 1996. The evolution of body mass and relative brain size in fossil hominids. *Journal of Human Evolution* 30 (3): 243–276.

Kay, R. F. 1985. Dental evidence for the diet of *Australopithecus. Annual Review of Anthropology* 14 (1): 315–341.

Kay, R. F., C. Ross, and B. A. Williams. 1997. Anthropoid origins. *Science* 275 (5301):797–804.

Keeley, L. H. 1974. Technique and methodology in microwear studies: A critical review. *World Archaeology* 5 (3): 323–336.

Kehoe, A. B. 2006. *The ghost dance: Ethnohistory and revitalization.* Long Grove, IL: Waveland Press.

Kiyono-Fuse, M. 2008. Use of wet hair to capture swarming termites by a chimpanzee in Mahale, Tanzania. *Pan Africa News* 15 (1): 8–12.

Klein, R. G. 1995. Anatomy, behavior, and modern human origins. *Journal of World Prehistory* 9 (2):167–198.

———. 2009. *The human career: Human biological and cultural origins.* Chicago: University of Chicago Press.

Klein, R. G., and D. W. Bird. 2016. Shellfishing and human evolution. *Journal of Anthropological Archaeology* 44:198–205.

Kneebone, E. 1991. Interpreting traditional culture as land management. In *Aboriginal involvement in parks and protected areas: Papers presented to a conference organised by the Johnstone Centre of Parks, Recreation and Heritage at Charles Sturt University, Albury, New South Wales, 22–24 July 1991*, ed. J. Birkhead, T. DeLacy, and L. Smith, 227–238. Canberra: Canberra Aboriginal Studies Press.

Kottak, P. C. 2006. *Anthropology: Exploration of human diversity*. New York: McGraw-Hill.

Kouřimská, L., and A. Adámková. 2016. Nutritional and sensory quality of edible insects. *NFS Journal* 4 (October): 22–26.

Lack, D. L. 1968. *Ecological adaptations for breeding in birds*. London: Methuen.

Larsen, C. S. 1995. Biological changes in human populations with agriculture. *Annual Review of Anthropology* 24 (1): 185–213.

Leakey, L. S. B., P. V. Tobias, and J. R. Napier. 1964. A new species of the genus *Homo* from Olduvai Gorge. *Nature* 4927: 7–9.

Leakey, M. D. 1971. *Olduvai Gorge*. Vol. 3, *Excavations in beds I & II, 1960–1963*. Cambridge, MA: Cambridge University Press.

Lee, R. B. 1979. *The !Kung San: Men, women, and work in a foraging society*. Cambridge: Cambridge University Press.

Lee, R. B., and I. Devore, eds. 1968. *Man the hunter*. New York: Routledge.

Lee-Thorp, J., J. F. Thackeray, and N. van der Merwe. 2000. The hunters and hunted revisited. *Journal of Human Evolution* 39 (6): 565–576.

Lee-Thorp, J., N. van der Merwe, and C. K. Brain. 1994. Diet of *Australopithecus robustus* at Swartkrans from stable carbon isotopic analysis. *Journal of Human Evolution* 27 (4): 361–372.

Leonard, W. R., and M. L. Robertson. 1994. Evolutionary perspectives on human nutrition: The influence of brain and body size on diet and metabolism. *American Journal of Human Biology* 6 (1): 77–88.

———. 1997. Comparative primate energetics and hominid evolution. *American Journal of Physical Anthropology* 102 (2): 265–281.

Lesnik, J. J. 2011. Bone tool texture analysis and the role of termites in the diet of South African hominids. *PaleoAnthropology*: 268–281.

———. 2014. Termites in the hominin diet: A meta-analysis of termite genera, species and castes as a dietary supplement for South African robust australopithecines. *Journal of Human Evolution* 71 (June): 94–104.

———. 2017. Not just a fallback food: Global patterns of insect consumption related to geography, not agriculture. *American Journal of Human Biology* 29 (4): e22976.

Lesnik, J., and J. F. Thackeray. 2006. Analysis of crown heights of equid dentition from a site in the Cradle of Humankind: Behavioural implications. *Annals of the Transvaal Museum* 43: 118-120.

———. 2007. The efficiency of stone and bone tools for opening termite mounds: Implications for hominid tool use at Swartkrans. *South African Journal of Science* 103 (9–10): 354–358.

Leung, W.-T. W., F. Busson, and C. Jardin. 1968. *Food composition table for use in Africa*. Rome: Food and Agriculture Organization.

Leutenegger, W., and J. T. Kelly. 1977. Relationship of sexual dimorphism in canine size and body size to social, behavioral, and ecological correlates in anthropoid primates. *Primates* 18 (1): 117–136.

Li, M.-L., L. Sheng-Mei, D. Wang, and R.-P. Liu. 2006. Fatty acid composition and contents of five insect species. *Chinese Bulletin of Entomology* 2. In Chinese.

Linton, S. 1971. Woman the gatherer: Male bias in anthropology. In *Women in cross-cultural perspectives*, ed. S. Jacobs, 9–21. Urbana: University of Illinois.

Lonsdorf, E. V. 2005. Sex differences in the development of termite-fishing skills in the wild chimpanzees, *Pan troglodytes schweinfurthii*, of Gombe National Park, Tanzania. *Animal Behaviour* 70 (3): 673–683.

Looy, H., F. V. Dunkel, and J. R. Wood. 2014. How then shall we eat? Insect-eating attitudes and sustainable foodways. *Agriculture and Human Values* 31 (1): 131–141.

Looy, H., and J. R. Wood. 2006. Attitudes toward invertebrates: Are educational "Bug Banquets" effective? *Journal of Environmental Education* 37 (2): 37–48.

Lovejoy, C. O. 1981. The origin of man. *Science* 211 (4480): 341–350.

———. 2009. Reexamining human origins in light of *Ardipithecus ramidus*. *Science* 326 (5949): 74–74e78.

Ludwig, F., de H. Kroon, and H. H. Prins. 2008. Impacts of savanna trees on forage quality for a large African herbivore. *Oecologia* 155 (3): 487–496.

Lukas, D., and T. H. Clutton-Brock. 2013. The evolution of social monogamy in mammals. *Science* 341 (6145): 526–530.

Lundy, M. E., and M. P. Parrella. 2015. Crickets are not a free lunch: Protein capture from scalable organic side-streams via high-density populations of *Acheta domesticus*. *PLoS ONE* 10 (4): e0118785.

Madsen, D. B., and D. N. Schmitt. 1998. Mass collecting and the diet breadth model: A Great Basin example. *Journal of Archaeological Science* 25 (5): 445–455.

Magill, C. R., G. M. Ashley, M. Domínguez-Rodrigo, and K. H. Freeman. 2016. Dietary options and behavior suggested by plant biomarker evidence in an early human habitat. *Proceedings of the National Academy of Sciences* 113 (11): 2874–2879.

Mann, A. 1972. Hominid and cultural origins. *Man* 7 (3): 366–379.

Mann, M. E. 2002. Little Ice Age. In *Encyclopedia of global environmental change*. Vol. 1, ed. M. C. McCracken, J. S. Perry, and T. Munn, 504–509. Rexcdale, Ont.: Wiley.

Mann, M. E., Z. Zhang, S. Rutherford, R. Bradley, M. Hughes, D. Shindell, C. Ammann, G. Faluvegi, and F. Ni. 2009. Global signatures and dynamical origins of the Little Ice Age and medieval climate anomaly. *Science* 326 (5957): 1256–1260.

Marean, C. W., M. Bar-Matthews, J. Bernatchez, E. Fisher, P. Goldberg, A. I. R. Herries, Z. Jacobs, A. Jerardino, P. Karkanas, T. Minichillo, P. J. Nilssen, E. Thompson, I. Watts, and H. M. Williams. 2007. Early human use of marine resources and pigment in South Africa during the Middle Pleistocene. *Nature* 449 (7164): 905–908.

Marlowe, F. W. 2003. A critical period for provisioning by Hadza men: Implications for pair bonding. *Evolution and Human Behavior* 24 (3): 217–229.

———. 2007. Hunting and gathering: The human sexual division of foraging labor. *Cross-Cultural Research* 41 (2): 170–195.

Marlowe, F. W., and J. C. Berbesque. 2009. Tubers as fallback foods and their impact on Hadza hunter-gatherers. *American Journal of Physical Anthropology* 140 (4): 751–758.

Marreiros, J. M., J. F. Gibaja Bao, and N. Ferreira Bicho, eds. 2016. *Use-wear and residue analysis in archaeology*. New York: Springer.

Martin, D. 2011. What do bugs taste like, anyway? *Huffington Post*, July 18. Accessed November 28, 2017. https://www.huffingtonpost.com/daniella-martin/what-do-bugs-taste-like-a_b_901775.html.

McGraw, W. S., A. E. Vick, and D. J. Daegling. 2011. Sex and age differences in the diet and ingestive behaviors of sooty mangabeys (*Cercocebus atys*) in the Tai Forest, Ivory Coast. *American Journal of Physical Anthropology* 144 (1): 140–153.

McGrew, W. C. 1979. Evolutionary implications of sex differences in chimpanzee predation and tool use. In *The great apes*, ed. D. Hamburg and E. McGown, 441–463. Berkeley: Society for the Study of Human Evolution.

———. 1981. The female chimpanzee as a human evolutionary prototype. In *Woman the gatherer*, ed. F. Dahlberg, 35–73. New Haven, CT: Yale University Press.

———. 1992. *Chimpanzee material culture: Implications for human evolution*. Cambridge: Cambridge University Press.

———. 2001. The other faunivory: Primate insectivory and early human diet. In *Meat-eating and human evolution*, ed. C. Stanford, 160–178. Oxford: Oxford University Press.

———. 2014. The "other faunivory" revisited: Insectivory in human and non-human primates and the evolution of human diet. *Journal of Human Evolution* 71 (June): 4–11.

McGrew, W. C., L. F. Marchant, M. M. Beuerlein, D. Vrancken, B. Fruth, and G. Hohmann. 2007. Prospects for bonobo insectivory: Lui Kotal, Democratic Republic of Congo. *International Journal of Primatology* 28 (6): 1237–1252.

McHenry, H. M., and K. Coffing. 2000. *Australopithecus* to *Homo*: Transformations in body and mind. *Annual Review of Anthropology* 29: 125–146.

Melin, A. D., L. M. Fedigan, H. C. Young, and S. Kawamura. 2010. Invertebrate foraging by Costa Rican capuchin monkeys: Testing predicted sex differences in relation to color vision variation. Paper presented at the 33rd Annual Meeting of the American Society of Primatologists, Louisville, KY, June 16–19.

Melin, A. D., H. C. Young, K. N. Mosdossy, and L. M. Fedigan. 2014. Seasonality, extractive foraging, and the evolution of primate sensorimotor intelligence. *Journal of Human Evolution* 71 (June): 77–86.

Melnick, M. 2014. Actually, there were many paleo diets. *Huffington Post*, December 17. Accessed November 28, 2017. http://www.huffingtonpost.com/2014/12/17/paleo-diet-debunked_n_6342382.html.

Mguni, S. 2006. Iconography of termites' nests and termites: Symbolic nuances of formlings in southern African San rock art. *Cambridge Archaeological Journal* 16 (1): 53–71.

Miller, E. M. 2014. Iron status and reproduction in US women: National Health and Nutrition Examination Survey 1999–2006. *PLoS ONE* 9 (11): e112216.

Milton, K. 1999a. A hypothesis to explain the role of meat-eating in human evolution. *Evolutionary Anthropology: Issues, News, and Reviews* 8 (1): 11–21.

———. 1999b. Nutritional characteristics of wild primate foods: Do the diets of our closest living relatives have lessons for us? *Nutrition* 15 (6): 488–498.

Mineka, S., and M. Cook. 1988. Social learning and the acquisition of snake fear in monkeys. In *Social learning: Psychological and biological perspectives*, ed. T. Zentall and G. Galef, 51–73. Hillsdale, NJ: Erlbaum.

Mintz, S. W., and C. M. Du Bois. 2002. The anthropology of food and eating. *Annual Review of Anthropology* 31: 99–119.

Mitani, J. C., and D. P. Watts. 2001. Why do chimpanzees hunt and share meat? *Animal Behaviour* 61 (5): 915–924.

Moggi-Cecchi, J., C. Menter, S. Boccone, and A. Keyser. 2010. Early hominin dental remains from the Plio-Pleistocene site of Drimolen, South Africa. *Journal of Human Evolution* 58 (5): 374–405.

Montague, M. J., T. R. Disotell, and A. Di Fiore. 2014. Population genetics, dispersal, and kinship among wild squirrel monkeys (*Saimiri sciureus macrodon*): Preferential association between closely related females and its implications for insect prey capture success. *International Journal of Primatology* 35 (1): 169–187.

Morell, V. 1993. Called "trimates," three bold women shaped their field. *Science* 260 (5106): 420–425.

Moscovice, L. R., M. H. Issa, K. L. Petrzelkova, N. S. Keuler, C. T. Snowdon, and M. A. Huffman. 2007. Fruit availability, chimpanzee diet, and grouping patterns on Rubondo Island, Tanzania. *American Journal of Primatology* 69 (5): 487–502.

Mullendore, W. C. 1941. *History of the United States Food Administration, 1917–19*. Stanford, CA: Stanford University Press.

Müller, A., J. Evans, C.L. R. Payne, and R. Roberts. 2016. Entomophagy and power. *Journal of Insects as Food and Feed* 2 (2): 121<n>136.

Murray, C. M., E. V. Lonsdorf, L. E. Eberly, and A. E. Pusey. 2009. Reproductive energetics in free-living female chimpanzees (*Pan troglodytes schweinfurthii*). *Behavioral Ecology* 20 (6): 1211–1216.

National Research Council. 2003. *Nutrient requirements of non-human primates*. 2nd ed. Washington, DC: National Academy Press.

Navarrete, A., C. P. van Schaik, and K. Isler. 2011. Energetics and the evolution of human brain size. *Nature* 480 (7375): 91–93.

Nekaris, K. A. I. 2005. Foraging behaviour of the slender loris (*Loris lydekkerianus lydekkerianus*): Implications for theories of primate origins. *Journal of Human Evolution* 49 (3): 289–300.

Netshifhefhe, S. R., E. C. Kunjeku, and F. D. Duncan. 2018. Human uses and indigenous knowledge of edible termites in Vhembe District, Limpopo Province, South Africa. *South African Journal of Science* 114 (1–2): 1–10.

Newcomer, M., R. Grace, and R. Unger-Hamilton. 1986. Investigating microwear polishes with blind tests. *Journal of Archaeological Science* 13 (3): 203–217.

Newton-Fisher, N. 1999. Termite eating and food sharing by male chimpanzees in the Budongo Forest, Uganda. *African Journal of Ecology* 37 (3): 369–371.

Ni, X., D. L. Gebo, M. Dagosto, J. Meng, P. Tafforeau, J. J. Flynn, and K. C. Beard. 2013. The oldest known primate skeleton and early haplorhine evolution. *Nature* 498 (7452): 60–64.

Niassy, S., K. Fiaboe, H. Affognon, K. Akutse, M. Tanga, and S. Ekesi. 2016. African indigenous knowledge on edible insects to guide research and policy. *Journal of Insects as Food and Feed* 2 (3): 161–170.

Nilssen, A. C., and J. O. Gjershaug. 1988. Reindeer warble fly larvae found in red deer. *Rangifer* 8 (1): 35–37.

Nissen, K. 1973. Analysis of human coprolites from Bamert Cave, Amador County, California. In *The archaeology of Bamert Cave, Amador County, California*, ed. R. F. Heizer. Berkeley: University of California Archaeological Research Facility.

Nonaka, K. 1996. Ethnoentomology of the Central Kalahari San. *African Study Monographs* 22: 29–46.

O'Dea, K., P. A. Jewell, A. Whiten, S. A. Altmann, S. S. Strickland, and O. T. Oftedal. 1991. Traditional diet and food preferences of Australian Aboriginal hunter-gatherers. *Philosophical Transactions of the Royal Society of London B: Biological Sciences* 334 (1270): 233–241.

O'Malley, R. C., and W. C. McGrew. 2014. Primates, insects, and insect resources. *Journal of Human Evolution* 71 (June): 1–3.

O'Malley, R. C., and M. L. Power. 2012. Nutritional composition of actual and potential insect prey for the Kasekela chimpanzees of Gombe National Park, Tanzania. *American Journal of Physical Anthropology* 149 (4): 493–503.

———. 2014. The energetic and nutritional yields from insectivory for Kasekela chimpanzees. *Journal of Human Evolution* 71 (June): 46–58.

O'Malley, R. C., M. A. Stanton, I. C. Gilby, E. V. Lonsdorf, A. Pusey, A. C. Markham, and C. M. Murray. 2016. Reproductive state and rank influence patterns of meat consumption in wild female chimpanzees (*Pan troglodytes schweinfurthii*). *Journal of Human Evolution* 90 (January): 16–28.

O'Malley, R. C., W. Wallauer, C. M. Murray, and J. Goodall. 2012. The appearance and spread of ant fishing among the Kasekela chimpanzees of Gombe: A possible case of intercommunity cultural transmission. *Current Anthropology* 53 (5): 650.

Olatunji, B. O., J. Haidt, D. McKay, and B. David. 2008. Core, animal reminder, and contamination disgust: Three kinds of disgust with distinct personality, behavioral, physiological, and clinical correlates. *Journal of Research in Personality* 42 (5): 1243–1259.

Omotoso, O. T., and C. O. Adedire. 2007. Nutrient composition, mineral content, and the solubility of the proteins of palm weevil, *Rhynchophorus phoenicis f.* (Coleoptera: Curculionidae). *Journal of Zhejiang University Science B* 8 (5): 318–322.

Onyeike, E. N., E. O. Ayalogu, and C. C. Okaraonye. 2005. Nutritive value of the larvae of raphia palm beetle (*Oryctes rhinoceros*) and weevil (*Rhynchophorus pheonicis*). *Journal of the Science of Food and Agriculture* 85 (11): 1822–1828.

Oonincx, D. G., J. van Itterbeeck, M. J. Heetkamp, H. van den Brand, J. J. van Loon, and A. van Huis. 2010. An exploration on greenhouse gas and ammonia production by insect species suitable for animal or human consumption. *PLoS ONE* 5 (12): e14445.

Panger, M. A., A. S. Brooks, B. G. Richmond, and B. Wood. 2002. Older than the Oldowan? Rethinking the emergence of hominin tool use. *Evolutionary Anthropology* 11 (6): 235–245.

Paoletti, M. G., E. Buscardo, and D. L. Dufour. 2000. Edible invertebrates among Amazonian Indians: A critical review of disappearing knowledge. *Environment, Development and Sustainability* 2 (3–4): 195–225.

Peters, C. R., and J. C. Vogel. 2005. Africa's wild C_4 plant foods and possible early hominid diets. *Journal of Human Evolution* 48 (3): 219–236.

Pickett, S. B., C. M. Bergey, and A. Di Fiore. 2012. A metagenomic study of primate insect diet diversity. *American Journal of Primatology* 74 (7): 622–631.

Piperno, D. R. 2006. *Phytoliths: A comprehensive guide for archaeologists and paleoecologists*. New York: Altamira Press.

Pontzer, H., J. R. Scott, D. Lordkipanidze, and P. S. Ungar. 2011. Dental microwear texture analysis and diet in the Dmanisi hominins. *Journal of Human Evolution* 61 (6): 683–687.

Prestwich, G. D. 1984. Defense mechanisms of termites. *Annual Review of Entomology* 29 (1): 201–232.

Pruetz, J. D. 2006. Feeding ecology of savanna chimpanzees (*Pan troglodytes verus*) at Fongoli, Senegal. In *Feeding ecology in apes and other primates: Ecological, physical and behavioral aspects*, ed. G. Hohmann, M. M. Robbins, and C. Boesch, 161–182. Cambridge: Cambridge University Press.

Prüfer, K., F. Racimo, N. Patterson, F. Jay, S. Sankararaman, S. Sawyer, A. Heinze, G. Renaud, P. H. Sudmant, C. de Filippo, H. Li, S. Mallick, M. Dannemann, Q. Fu, M. Kircher, M. Kuhlwilm, M. Lachmann, M. Meyer, M. Ongyerth, M. Siebauer, C. Theunert, A. Tandon, P. Moorjani, J. Pickrell, J. C. Mullikin, S. H. Vohr, R. E. Green, I. Hellmann, P. L. Johnson, H. Blanche, H. Cann, J. O. Kitzman, J. Shendure, E. E. Eichler, E. S. Lein, T. E. Bakken, L. V. Golovanova, V. B. Doronichev, M. V. Shunkov, A. P. Derevianko, B. Viola, M. Slatkin, D. Reich, J. Kelso, and S. Pääbo. 2014. The complete genome sequence of a Neanderthal from the Altai Mountains. *Nature* 505 (7481): 43–49.

Rafert, J., and E. O. Vineberg. 1997. Bonobo nutrition—relation of captive diet to wild diet. In *Bonobo Husbandry Manual*, 3.1–3.18. American Association of Zoos and Aquariums. Accessed February 8, 2018. https://nagonline.net/wp-content/uploads/2013/12/Bonobo-Nutrition1.pdf.

Rakison, D. H., and J. Derringer. 2008. Do infants possess an evolved spider-detection mechanism? *Cognition* 107 (1): 381–393.

Ramos-Elorduy, J., J. M. P. Moreno. E. E. Prado, M. A. Perez, J. L. Otero, and O. L. De Guevara. 1997. Nutritional value of edible insects from the state of Oaxaca, Mexico. *Journal of Food Composition and Analysis* 10 (2): 142–157.

Raubenheimer, D., and J. M. Rothman. 2013. Nutritional ecology of entomophagy in humans and other primates. *Annual Review of Entomology* 58 (1): 141–160.

Raubenheimer, D., J. M. Rothman, H. Pontzer, and S. J. Simpson. 2014. Macronutrient contributions of insects to the diets of hunter–gatherers: A geometric analysis. *Journal of Human Evolution* 71: 70<n>76.

Redford, K. H., and J. G. Dorea. 1984. The nutritional value of invertebrates with emphasis on ants and termites as food for mammals. *Journal of Zoology* 203 (3): 385–395.

Richards, M. P., P. B. Pettitt, M. C. Stiner, and E. Trinkaus. 2001. Stable isotope evidence for increasing dietary breadth in the European mid-Upper Paleolithic. *Proceedings of the National Academy of Sciences* 98 (11): 6528–6532.

Richards, M. P., P. B. Pettitt, E. Trinkaus, F. H. Smith, M. Paunović, and I. Karavanić. 2000. Neanderthal diet at Vindija and Neanderthal predation: The evidence from stable isotopes. *Proceedings of the National Academy of Sciences* 97 (13): 7663–7666.

Richards, M. P., and R. W. Schmitz. 2008. Isotope evidence for the diet of the Neanderthal type specimen. *Antiquity* 82 (317): 553–559.

Richards, M. P., and E. Trinkaus. 2009. Isotopic evidence for the diets of European Neanderthals and early modern humans. *Proceedings of the National Academy of Sciences* 106 (38): 16034–16039.

Richardson, P. R. K. 1987. Aardwolf: The most highly specialised myrmecophagous mammal? *South African Journal of Science* 83: 405–410.

Rose, L., and F. Marshall. 1996. Meat eating, hominid sociality, and home bases revisited. *Current Anthropology* 37 (2): 307–338.

Rothman, J. M., D. Raubenheimer, M. A. Bryer, M. Takahashi, and C. C. Gilbert. 2014. Nutritional contributions of insects to primate diets: Implications for primate evolution. *Journal of Human Evolution* 71 (June): 59–69.

Rowe, N. 1996. *Pictorial guide to the living primates.* Charlestown, RI: Pogonias Press.

Rozin, P., J. Haidt, and C. R. McCauley. 1999. Disgust: The body and soul emotion. In *Handbook of cognition and emotion*, ed. Tim Dalgleish and Mick Power, 429–445. New York: John Wiley & Sons.

Ruddle, K. 1973. The human use of insects: Examples from the Yukpa. *Biotropica* 5 (2): 94–101.

Rumpold, B. A., and O. K. Schlüter. 2013. Nutritional composition and safety aspects of edible insects. *Molecular Nutrition & Food Research* 57 (5): 802–823.

Sabrosky, C. W. 1953. How many insects are there? *Systematic Zoology* 2 (1): 31–36.

Sands, W. A. 1965. Mound population movements and fluctuations in *Trinervitermes ebenerianus Sjöstedt* (Isoptera, Termitidae, Nasutitertinae). *Insectes Sociaux* 12 (1): 49–58.

Sanz, C., D. Morgan, and S. Gulick. 2004. New insights into chimpanzees, tools, and termites from the Congo Basin. *The American Naturalist* 154 (5): 567–581.

Sanz, C. M., C. Schöning, and D. B. Morgan. 2010. Chimpanzees prey on army ants with specialized tool set. *American Journal of Primatology* 72 (1): 17–24.

Savadogo, P., M. Tigabu, L. Sawadogo, and P. C. Odén. 2009. Herbaceous phytomass and nutrient concentrations of four grass species in Sudanian savanna woodland subjected to recurrent early fire. *African Journal of Ecology* 47 (4): 699–710.

Saxe, G. A., J. M. Major, J. Y. Nguyen, K. M. Freeman, T. M. Downs, and C. E. Salem. 2006. Potential attenuation of disease progression in recurrent prostate cancer with plant-based diet and stress reduction. *Integrative Cancer Therapies* 5 (3): 206–213.

Schoeninger, M. J. 2014. Stable isotope analyses and the evolution of human diets. *Annual Review of Anthropology* 43: 413–430.

Schoeninger, M. J., H. T. Bunn, S. S. Murray, and J. A. Marlett. 2001. Composition of tubers used by Hadza foragers of Tanzania. *Journal of Food Composition and Analysis* 14 (1): 15–25.

Schoeninger, M. J., U. T. Iwaniec, and L. T. Nash. 1998. Ecological attributes recorded in stable isotope ratios of arboreal prosimian hair. *Oecologia* 113 (2): 222–230.

Schöning, C., T. Humle, Y. Möbius, and W. C. McGrew. 2008. The nature of culture: Technological variation in chimpanzee predation on army ants revisited. *Journal of Human Evolution* 55 (1): 48–59.

Semaw, S. 2000. The world's oldest stone artefacts from Gona, Ethiopia: Their implications for understanding stone technology and patterns of human evolution between 2.6–1.5 million years ago. *Journal of Archaeological Science* 27 (12): 1197–1214.

Shao-jun, W. Z.-Q. C., and C. Yong-huang. 1997. Nutritive value of *Herse convolvuli* (Lep.: Sphingidae). *Wuyi Science Journal* 13: 1997–1900.

Shen, G., X. Gao, B. Gao, and D. E. Granger. 2009. Age of Zhoukoudian *Homo erectus* determined with 26 Al/10 Be burial dating. *Nature* 458:198-200.

Shockley, M., and A. T. Dossey. 2014. Insects for human consumption. In *Mass production of beneficial organisms*, ed. J. A. Morales-Mores, M. Guadalupe Rojas, and D. I. Shapiro-Ilan, 617–652. San Diego: Academic Press.

Simoons, F. J. 1961. *Eat not this flesh: Food avoidances in the Old World*. Madison, WI: University of Wisconsin Press.

Simoons, F. J., S. M. Batra, A. K. Chakravarti, P. Diener, G. E. Ferro-Luzzi, M. Harris, A. Heston, R. Hoffpauir, D. O. Lodrick, S. L. Malik, W. E. Mey, S. N. Mishra, S. Odend'hal, R. P. Palmieri, D. Pimentel, E. E. Robkin, C. W. Schwabe, J. E. Schwartzberg, M. Suryanarayana, and P. L. Wagner. 1979. Questions in the sacred-cow controversy [and comments and reply]. *Current Anthropology* 20 (3): 467–493.

Sistiaga, A., C. Mallol, B. Galván, and R. E. Summons. 2014. The Neanderthal meal: A new perspective using faecal biomarkers. *PLoS ONE* 9 (6): e101045.

Smaby, B. P. 1975. The Mormons and the Indians: Conflicting ecological systems in the Great Basin. *American Studies* 16 (1): 35–48.

Smith, R. J., and W. L. Jungers. 1997. Body mass in comparative primatology. *Journal of Human Evolution* 32 (6): 523–559.

Sogari, G. 2015. Entomophagy and Italian consumers: An exploratory analysis. *Progress in Nutrition* 17 (4): 311–316.

Solodenko, N., A. Zupancich, S. N. Cesaro, O. Marder, C. Lemorini, and R. Barkai. 2015. Fat residue and use-wear found on Acheulian biface and scraper associated with butchered elephant remains at the site of Revadim, Israel. *PLoS ONE* 10 (3): e0118572.

Sponheimer, M., and J. Lee-Thorp. 1999. Isotopic evidence for the diet of an early hominid, *Australopithecus africanus*. *Science* 283 (5400): 368–370.

Sponheimer, M., J. Lee-Thorp, D. de Ruiter, D. Codron, J. Codron, A. T. Baugh, and F. Thackeray. 2005. Hominins, sedges, and termites: New carbon isotope data from the Sterkfontein Valley and Kruger National Park. *Journal of Human Evolution* 48 (3): 301–312.

Sponheimer, M., B. H. Passey, D. J. De Ruiter, D. Guatelli-Steinberg, T. E. Cerling, and J. A. Lee-Thorp. 2006. Isotopic evidence for dietary variability in the early hominin *Paranthropus robustus*. *Science* 314 (5801): 980–982.

Srivastava, S., N. Babu, and H. Pandey. 2009. Traditional insect bioprospecting—As human food and medicine. *Indian Journal of Traditional Knowledge* 8 (4): 485–494.

Stahl, A. B. 1984. Hominid dietary selection before fire. *Current Anthropology* 25 (2): 151–168.

Stanford, C., J. S. Allen, and S. C. Antón. 2011. *Biological anthropology: The natural history of humankind*. Upper Saddle River, NJ: Pearson Higher Education.

———. 2013. *Biological anthropology*. 3rd ed. New York: Pearson.

Stanhope, M. J., V. G. Waddell, O. Madsen, W. De Jong, S. B. Hedges, G. C. Cleven, D. Kao, and M. S. Springer. 1998. Molecular evidence for multiple origins of Insectivora and for a new order of endemic African insectivore mammals. *Proceedings of the National Academy of Sciences* 95 (17): 9967–9972.

Steinfeld, H., P. Gerber, T. Wassenaar, V. Castel, M. Rosales, and C. de Haan. 2006. *Livestock's long shadow: Environmental issues and options*. Rome: Food and Agriculture Organization of the United Nations.

Stemp, W. J., A. S. Watson, and A. A. Evans. 2015. Surface analysis of stone and bone tools. *Surface Topography: Metrology and Properties* 4 (1): 013001.

Strait, S. G. 2014. Myrmecophagous microwear: Implications for diet in the hominin fossil record. *Journal of Human Evolution* 71 (June): 87–93.

Strier, K. B. 2015. *Primate behavioral ecology*. New York: Routledge.

Sussman, R. W. 1991. Primate origins and the evolution of angiosperms. *American Journal of Primatology* 23 (4): 209–223.

Sussman, R. W., D. T. Rasmussen, and P. H. Raven. 2013. Rethinking primate origins again. *American Journal of Primatology* 75 (2): 95–106.

Sutton, M. Q. 1988. *Insects as food: Aboriginal entomophagy in the Great Basin*. Menlo Park, NJ: Ballena Press.

———. 1995. Archaeological aspects of insect use. *Journal of Archaeological Method and Theory* 2 (3): 253–298.

Sweeney, G. 1947. Food supplies of a desert tribe. *Oceania* 17 (4): 289–299.

Takeda, J. 1990. The dietary repertory of the Ngandu people of the tropical rain forest: An ecological and anthropological study of the subsistence activities and food procurement technology of a slash-and-burn agriculturist in the Zaire river basin. *African Study Monographs* 11 (Suppl.): 1–75.

Tanner, N., and A. Zihlman. 1976. Women in evolution. Part I: Innovation and selection in human origins. *Signs: Journal of Women in Culture and Society* 1 (3): 585–608.

Tattersall, I. 2009. Human origins: Out of Africa. *Proceedings of the National Academy of Sciences* 106 (38): 16018–16021.

Taylor, W. A., P. A. Lindsey, and J. D. Skinner. 2002. The feeding ecology of the aardvark *Orycteropus afer*. *Journal of Arid Environments* 50 (1): 135–152.

Thorne, A., R. Grün, G. Mortimer, N. A. Spooner, J. J. Simpson, M. McCulloch, L. Taylor, and D. Curnoe. 1999. Australia's oldest human remains: Age of the Lake Mungo 3 skeleton. *Journal of Human Evolution* 36 (6): 591–612.

Thrussell, E. 2016. A recipe for identity: Food and culture in Oaxaca, Mexico. PhD diss., University of Adelaide.

Towle, I., D. Irish, and I. De Groote. 2017. Behavioral inferences from the high levels of dental chipping in *Homo naledi*. *American Journal of Physical Anthropology* 164 184–192.

Truswell, A. S. 1977. Diet and nutrition of hunter-gatherers. In *Health and disease in tribal societies*, ed. K. M. Elliot and J. Whelan, 213–221. Amsterdam: Elsevier.

Tuso, P. J., M. H. Ismail, B. P. Ha, and C. Bartolotto. 2013. Nutritional update for physicians: Plant-based diets. *The Permanente Journal* 17 (2): 61.

Tutin, C. E. G., and M. Fernandez. 1992. Insect-eating by sympatric lowland gorillas (*Gorilla g. gorilla*) and chimpanzees (*Pan t. troglodytes*) in the Lopé Reserve, Gabon. *American Journal of Primatology* 28 (1): 29–40.

Tylor, E. B. 1871. *Primitive culture: Researches into the development of mythology, philosophy, religion, art, and custom*. Vol. 2. London: J. Murray.

Uehara, S. 1982. Seasonal changes in the techniques employed by wild chimpanzees in the Mahale Mountains, Tanzania, to feed on termites (*Pseudacanthotermes spiniger*). *Folia Primatologica* 37 (1–2): 44–76.

———. 1986. Sex and group differences in feeding on animals by wild chimpanzees in the Mahale Mountains National Park, Tanzania. *Primates* 27 (1): 1–13.

Ungar, P. S. 2007. *Evolution of the human diet: The known, the unknown, and the unknowable*. Oxford: Oxford University Press.

Ungar, P. S., K. L. Krueger, R. J. Blumenschine, J. Njau, and R. S. Scott. 2012. Dental microwear texture analysis of hominins recovered by the Olduvai Landscape Paleoanthropology Project 1995–2007. *Journal of Human Evolution* 63 (2): 429–437.

Ungar, P. S., and M. Sponheimer. 2011. The diets of early hominins. *Science* 334 (6053): 190–193.

Väisänen, R., and K. Heliövaara. 1994. Hot-spots of insect diversity in northern Europe. *Annales Zoologici Fennici* 31 (1): 71–81.

van der Merwe, N. J., J. F. Thackeray, J. A. Lee-Thorp, and J. Luyt. 2003. The carbon isotope ecology and diet of *Australopithecus africanus* at Sterkfontein, South Africa. *Journal of Human Evolution* 44 (5): 581–597.

van der Merwe, N. J., F. T. Masao, and M. K. Bamford. 2008. Isotopic evidence for contrasting diets of early hominins *Homo habilis* and *Australopithecus boisei* of Tanzania. *South African Journal of Science* 104 (3–4): 153–155.

Van Huis, A., J. Van Itterbeeck, H. Klunder, E. Mertens, A. Halloran, G. Muir, and P. Vantomme. 2013. *Edible insects: Future prospects for food and feed security*. Rome: Food and Agriculture Organization of the United Nations.

van Ryneveld, K. 2003. Bone tools from the early hominid sites, Guateng: An experimental approach. MS thesis, University of Witswatersrand.

Verbeke, W. 2015. Profiling consumers who are ready to adopt insects as a meat substitute in a western society. *Food Quality and Preference* 39 (January): 147–155.

Vincent, A. S. 1985. Plant foods in savanna environments: A preliminary report of tubers eaten by the Hadza of northern Tanzania. *World Archaeology* 17 (2): 131–148.

Vrba, E. S. 1975. Some evidence of chronology and palaeoecology of Sterkfontein, Swartkrans, and Kromdraai from the fossil Bovidae. *Nature* 254 (March 27): 301–304.

Wakayama, E. J., J. W. Dillwith, R. W. Howard, and G. J. Blomquist. 1984. Vitamin B_{12} levels in selected insects. *Insect Biochemistry* 14 (2): 175–179.

Walrath, D. 2006. Gender, genes, and the evolution of human birth. In *Feminist anthropology: Past present and future*, ed. P. L. Geller and M. K. Stockett, 55–69. Philadelphia: University of Pennsylvania Press.

Watts, D. P. 1989. Ant eating behavior of mountain gorillas. *Primates* 30 (1): 121–125.

White, F. J. 1992. Activity budgets, feeding behaviour and habitat use of pygmy chimpanzees at Lomako, Zaire. *American Journal of Primatology* 26 (3): 215–223.

White, T. D., B. Asfaw, Y. Beyene, Y. Haile-Selassie, C. O. Lovejoy, G. Suwa, and G. WoldeGabriel. 2009. *Ardipithecus ramidus* and the paleobiology of early hominids. *Science* 326 (5949): 64–86.

Whiten, A., J. Goodall, W. C. McGrew, T. Nishida, V. Reynolds, Y. Sugiyama, C. E. G. Tutin, R. W. Wrangham, and C. Boesch. 1999. Cultures in chimpanzees. *Nature* 399 (6737): 682–685.

Wilk, R. 2006. "But the young men don't want to farm any more": Political ecology and consumer culture in Belize. In *Reimagining political ecology*, ed. A. Biersack and J. B. Greenberg, 149–170. Durham, NC: Duke University Press.

Wolpoff, M. H., J. Hawks, D. W. Frayer, and K. Hunley. 2001. Modern human ancestry at the peripheries: A test of the replacement theory. *Science* 291 (5502): 293–297.

Wrangham, R., and D. Pilbeam. 2001. African apes as time machines. In *All apes great and small.* Vol. 1, *African apes*, ed. B. M. F. Galdikis, N. E. Briggs, L. K. Sheeran, G. L. Shapiro, and J. Goodall, 5–17. New York: Kluwer Academic.

Wrangham, R. W., N. L. Conklin, C. A. Chapman, and K. D. Hunt. 1991. The significance of fibrous foods for Kibale Forest chimpanzees. *Philosophical Transactions of the Royal Society of London B: Biological Sciences* 334 (1270): 171<n>178.

Wrangham, R. W., J. H. Jones, G. Laden, D. Pilbeam, and N. Conklin-Brittain. 1999. The raw and the stolen: Cooking and the ecology of human origins. *Current Anthropology* 40 (5): 567–577.

Wynn, T., and W. C. McGrew. 2001. An ape's view of the Oldowan. *Man* 24 (3): 383–398.

Yamagiwa, J., N. Mwanza, T. Yumoto, and T. Maruhashi. 1991. Ant eating by eastern lowland gorillas. *Primates* 32 (2): 247–253.

Yen, A., C. Bilney, M. Shackleton, and S. Lawler. 2017. Current issues involved with the identification and nutritional value of wood grubs consumed by Australian Aborigines. *Insect Science* (March 14): 1–12.

Zihlman, A. L. 1978. Women in evolution, part II: Subsistence and social organization among early hominids. *Signs: Journal of Women in Culture and Society* 4 (1): 4–20.

Zihlman, A., and N. Tanner. 1978. Gathering and the hominid adaptation. In *Female hierarchies*, ed. L. Tiger and H. Fowler, 163–194. London: Transaction Publishers.

Zimmerman, P. R., J. P. Greenberg, S. O. Wandiga, and P. J. Crutzen. 1982. Termites: A potentially large source of atmospheric methane, carbon dioxide, and molecular hydrogen. *Science* 218 (4572): 563–565.

INDEX

Page numbers in *italics* followed by *f, m,* and *t* indicate figures, maps, and tables.

Julie J. Lesnik is assistant professor of anthropology at Wayne State University. She is a 2015–2016 recipient of the American Fellowship from the American Association of University Women and a 2017–2018 fellow of the Leshner Leadership Institute for Public Engagement with Science of the American Association for the Advancement of Science.

CPSIA information can be obtained
at www.ICGtesting.com
Printed in the USA
LVHW04*2226010618
579330LV00002B/6/P

9 780813 056999